GROUP
THEORETIC
CRYPTOGRAPHY

CHAPMAN & HALL/CRC
CRYPTOGRAPHY AND NETWORK SECURITY

Series Editors

Douglas R. Stinson and Jonathan Katz

Published Titles

CHAPMAN & HALL/CRC
CRYPTOGRAPHY AND NETWORK SECURITY

GROUP THEORETIC CRYPTOGRAPHY

María Isabel González Vasco

Universidad Rey Juan Carlos
Madrid, Spain

Rainer Steinwandt

Florida Atlantic University
Boca Raton, FL

CRC Press
Taylor & Francis Group
Boca Raton London New York

CRC Press is an imprint of the
Taylor & Francis Group an **informa** business

A CHAPMAN & HALL BOOK

CRC Press
Taylor & Francis Group
6000 Broken Sound Parkway NW, Suite 300
Boca Raton, FL 33487-2742

© 2015 by Taylor & Francis Group, LLC
CRC Press is an imprint of Taylor & Francis Group, an Informa business

No claim to original U.S. Government works

Printed on acid-free paper
Version Date: 20150227

International Standard Book Number-13: 978-1-58488-836-9 (Hardback)

Visit the Taylor & Francis Web site at
http://www.taylorandfrancis.com

and the CRC Press Web site at
http://www.crcpress.com

To our families

Contents

List of Figures

Symbol Description

$\leftarrow$	uniformly at random selection from a given set		quotient group, ring, vector space, etc.
$\circ$	functional composition	$B \trianglelefteq A$	the subset B is an ideal of the ring A, or the subgroup B is normal in the group A
A^*	sequences formed with elements from an alphabet set A, or set of units of an algebraic structure A		
		B_n	braid group on n strands
A^+	non-empty sequences formed with elements from an alphabet set A	$|G : H|$	the index of H in G, for H a substructure of G— group, vector space, etc.
$\mathrm{Aut}(A)$	automorphisms of A	$\gcd(a, b)$	greatest common divisor of the integers a and b
ϵ	the identity element of a multiplicative algebraic structure		
		$\mathrm{Hom}(G, H)$	homomorphisms from G to H
$\cap$	set intersection		
$\cup$	set union	$\left(\frac{a}{b}\right)$	Jacobi or Legendre symbol
$\backslash$	set difference		
$X \times Y$	the cartesian product of sets or algebraic structures X and Y	$\ker(\cdot)$	kernel of a homomorphism
		$\inf(b)$	infimum of the braid b
$|A|$	cardinality of set A or order of an algebraic structure (generated by) A, or length of an element	LB_n	left braids in the braid group on n strands
		$\ln n$	natural logarithm of the positive real number n
$A \subseteq B$	A is a subset of B	RB_n	right braids in the braid group on n strands
$\emptyset$	empty set		
$\mathbb{F}_q$	finite field with q elements	$\mathbb{N}$	set of positive integers $1, 2, 3, \ldots$
$\mathrm{GL}_n(R)$	general linear group over the ring R	$\mathbb{N}_0$	set of non-negative integers $0, 1, 2, \ldots$
		$\mathbb{R}$	set of real numbers
$B \leq A$	the subset B is a substructure of A— subgroup, subring, vector subspace, etc.	$\mathbb{R}_{\geq 0}$	set of non-negative real numbers
		$\mathrm{SL}_n(R)$	special linear group over a ring R
A/B	the quotient structure subset B induces in A—	$\mathrm{Stab}(L)$	stabilizer of the set L under a group action
		$\mathbb{Z}$	set of integers

Preface

Group theory is one of the most beautiful areas of mathematics. Being the algebraic language of symmetries, it finds applications in an enormous variety of fields, not only within mathematics but notably within physics, chemistry, and life sciences. Group-theoretic techniques have propelled many scientific achievements from understanding molecular systems biology to analyzing strategies to solve Rubik's cube.

This book was conceived as an introduction to cryptography for readers lured by group theory who are curious to understand possible interplays between the two fields. Even though many constructions within modern cryptography exploit certain properties of finite groups, it is fair to say they are mostly number theoretic in nature. Only a few constructions make use of advanced group-theoretic tools. Notwithstanding this, the history of non-Abelian groups in cryptography dates back several decades. In the late 1970s, Magliveras explored the use of *permutation group mappings* for constructing a block cipher. In the early days of public-key cryptography, an asymmetric encryption scheme exploiting the hardness of the word problem in finitely presented groups was presented in 1984 at the Crypto conference in Santa Barbara, California, one of the flagship conferences in the field. Ever since then there have been numerous attempts to exploit hard group-theoretic problems for cryptographic constructions, many inspiring ideas have come to light and clever cryptanalytic strategies have shown that coming up with practical designs is a challenging target. At the same time, interesting theoretical questions within group theory have emerged from the cryptographic designs. Differing from what happens within "standard" computational group theory, the computational problems arising from the cryptographic perspective set up a scenario in which mathematical reasoning is challenged by computational limitations and, moreover, by the hardness of formally capturing the behavior of malicious entities in real-life applications.

For a smooth reading of this text some elementary mathematical maturity is assumed. Namely, the reader should be familiar with material commonly covered in undergraduate courses on linear algebra and discrete mathematics. We also tacitly assume elementary knowledge of algorithms. We have made an effort to make the text accessible for both computer science and mathematics undergraduates who have had some exposure to college-level mathematics and algorithms courses.

The book is divided into five separate parts. The first two chapters in Part I are devoted to reviewing the basic background knowledge from group theory and complexity theory. They are very short and intended as a minimalist review of the notions needed to undergo the rest of the text. Similarly, the third chapter offers a brief informal introduction to cryptology for those who are first approaching the field. Again, it is by no means a thorough introduction, but the topics and examples have been chosen in the hope that the unexperienced reader will get a historical glimpse of this quickly evolving area. Part II of the book is devoted to public-key encryption, a cryptographic subject which receives a lot of attention from the mathematical community. Even though understanding the basics of public-key encryption may help in penetrating other cryptographic constructions, this part is to a large extent independent of the subsequent two parts of the book. In particular, the techniques in the third part, devoted to secret-key encryption, are quite different. Sometimes not much group theory is brought to use in this field, but from a cryptographic point of view, familiarity with a block cipher such as the Advanced Encryption Standard is certainly relevant. Readers who are mainly interested in seeing group-theoretic constructions may at least want to take a look at Permutation Group Mappings in Section 7.3 and Cayley hash functions discussed in Section 8.3. The last part of the book, before the appendix, discusses digital signatures, one of the most important applications of cryptography next to encryption, as well as identification protocols. We also include a discussion of key establishment protocols, where proposals using non-Abelian groups—most notably braid groups—have been made. The Anshel–Anshel–Goldfeld protocol discussed in Section 9.2 received a lot of attention in the literature.

Each chapter concludes with a short section summarizing its content and providing references for further reading on the topic treated. Furthermore, we included a list of exercises, and solutions for some of them are provided in the appendix at the end of the book.

Acknowledgments

This text benefited tremendously from the interaction with mathematics and computer science students, each of them approaching the world of cryptography from their own unique perspective. Our views on the subject certainly have been influenced and shaped by the interaction with many colleagues, who generously shared their expertise in cryptography and group theory with us. We thank each and all of them and want to mention at least some of them explicitly: Ronald Cramer, Juan González-Meneses, Consuelo Martínez, Adriana Suárez Corona, Tran van Trung, Boaz Tsaban, and Jorge Luis Villar.

Very special thanks go to Thomas Beth, who brought us together, but sadly is no longer with us today, to Spyros Magliveras, who helped to initiate this project, and to Igor Shparlinski, for helping to unearth the beautiful mathematics within cryptography. To the team at CRC Press we owe a big thank you for being extremely helpful and patient with us. Last and not least we are indebted to our families and partners for their unconditional support and always bearing with us. Thank you Carlos and María, thank you Leigh, thank you Nacho. A heartfelt thank you goes to Palmira, who will sadly not see the outcome of this work.

Part I

Preliminaries

Chapter 1

Mathematical background

Let us start by reviewing some basic mathematical definitions and facts which are needed for smoothly going through this book. Our approach here is rather minimalistic in nature, and those who are willing to know a bit more mathematics (which will indeed be helpful) are directed to a number of texts and readings at the end of this chapter.

1.1 Algebraic structures in a nutshell

Throughout we assume familiarity with basic set-theoretic concepts and use standard notation for operations (union, intersection, difference, cartesian product) and terminology (subset, empty set, cardinality). When needed, the symbol description at the beginning of the book can serve as a reminder of common notations.

DEFINITION 1.1 [Equivalence relation]   *Let X be a set and R a subset of $X \times X$. Given $x, y \in X$, let us write xRy instead of $(x, y) \in R$. We say R is an* equivalence relation *on X provided that all of the following conditions hold:*

i. xRx for all $x \in X$ (reflexivity),

ii. for all $x, y \in X$ if xRy, then yRx (symmetry),

iii. for all $x, y, z \in X$, if xRy and yRz, then xRz (transitivity).

An equivalence relation on a set X induces a partition of X, i.e., a decomposition of X into disjoint subsets, each consisting of related elements. We refer to these subsets as *equivalence classes*. When R is clear from the context, it is common to simply write $[x]$ for the equivalence class

$$[x] = \{y \in X \mid xRy\}$$

of element x. The set comprised of all equivalence classes is called *quotient set of X by R* and denoted by X/R.

A *function* from a set X to a set Y is a law assigning an element $f(x)$ from Y to each element of X, and we refer to X as the *domain* of f, while the set $\{y \in Y \mid \exists x \in X \text{ such that } f(x) = y\}$ is called the *image* of f and denoted by $f(X)$. The sets we are concerned with in this book usually come along with some algebraic structure.

DEFINITION 1.2 [Semigroup]   *A semigroup is a pair $(S, \cdot)$ consisting of a non-empty set S and an associative binary operation $\cdot$ defined on S; that is, $\cdot$ is a function from $S \times S$ to S fulfilling: $s \cdot (t \cdot u) = (s \cdot t) \cdot u$ for all $s, t, u \in S$.*

Many semigroups come with a neutral element, and such semigroups are commonly known as *monoids*.

DEFINITION 1.3 [Monoid]   *A semigroup which has a distinguished element $\epsilon \in S$, such that $s \cdot \epsilon = \epsilon \cdot s$ for all $s \in X$, is called a monoid. The element ϵ is called the identity of S.*

Example 1.1
The set of non-negative integers is an additive monoid, with identity element 0.
$\square$

Let S be a monoid. An element $s \in S$ is called a *unit* if there exists an element $t \in S$ which satisfies

$$s \cdot t = t \cdot s = \epsilon.$$

Usually, t is denoted as s^{-1} and is called the *inverse* of s. It is customary to denote by S^* the set of all units in the monoid S.

Example 1.2
For any natural number $n \in \mathbb{N}$, denote by $M_n(\mathbb{R})$ the set of $n \times n$ matrices with real-valued entries. Considering standard matrix multiplication, $M_n(\mathbb{R})$ is a monoid—the identity element is the $n \times n$ identity matrix I_n (i.e., the $n \times n$ matrix with diagonal entries equal to one and all other entries being zero).
$\square$

In this book, many monoids S we are dealing with satisfy $S = S^*$, i.e., they are a group.

DEFINITION 1.4 [Group]   *A group is a monoid S in which every element is a unit, i.e., a monoid that satisfies $S = S^*$.*

In an abuse of notation, it is customary to write S instead of $(S, \cdot)$ for referring to the semigroup/monoid/group structure, but some caution is appropriate when doing so.

Example 1.3
The set of integers $\mathbb{Z}$ is a group under standard addition, but it is not under multiplication (as the only multiplicative units are 1 and -1). ▯

Each of the above algebraic structures is said to be *Abelian* if it fulfills the *commutative property*; namely, for any given $s, t \in S$ the equality $s \cdot t = t \cdot s$ holds. One often writes st or $s + t$ instead of $s \cdot t$ using, respectively, *multiplicative* or *additive* notation, at convenience. Additive notation is mainly common when dealing with a commutative operation. In multiplicative notation, the identity element is often denoted by 1 or ϵ. Analogously, in additive notation it is common to write 0 for the identity. Following common convention, we will write $|S|$ to denote the number of elements in a finite algebraic structure S. Especially in the context of finite groups, $|S|$ is also called *order* of S.

Given a certain algebraic structure, it is always of interest to study subsets of it that preserve the structure.

DEFINITION 1.5 [Subgroup] *A subgroup of a given group G is a non-empty subset $H \subseteq G$ which inherits the binary operation $\cdot$ from G, i. e., for any $g, h \in H$, we have $g \cdot h^{-1} \in H$. The fact that H is a subgroup of G is expressed as $H \leq G$.*

Every group G with $|G| \geq 2$ has at least two different subgroups: the group itself, and the one consisting only of the identity element (the latter is often denoted by 1). Any other subgroup of G is said to be *proper*.

Example 1.4
Every subgroup of $(\mathbb{Z}, +)$ consists of the multiples of a fixed non-negative integer n. Such subgroup is usually denoted by $n\mathbb{Z}$. ▯

Given a set of group elements $S \subseteq G$ one may consider the minimal subgroup of G containing S. This is called the *subgroup of G generated by S* and commonly written as $\langle S \rangle$. For a group element $g \in G$, its order, denoted by $|g|$, is defined as the order of the subgroup $\langle g \rangle$. If $|g|$ is finite, this number coincides with the smallest integer m fulfilling $g^m = \epsilon$; namely, if $g^k = \epsilon$, then k is a multiple of the order of g. Whenever $\langle g \rangle = G$, then g is called a *generator* of G and we say that G is *cyclic*, as all elements in G are powers of the generator g. Cyclic groups play a crucial role in public-key cryptography. The algorithmic task of actually finding a representation of a uniformly at random chosen element in a (large) finite cyclic group in terms of a given generator is at the foundation of many cryptographic constructions.

Given two group elements g, h from a certain group G, they are called *conjugate* if there exists a group element $m \in G$ such that $h = m^{-1}gm$. Subgroups that contain the conjugates of its elements are of special relevance:

DEFINITION 1.6 [Normal subgroup]   *Let G be a group and $H \leq G$ a subgroup. Then the subgroup H is said to be* normal, *denoted by $H \trianglelefteq G$, if $g^{-1}hg \in H$ for all $h \in H, g \in G$.*

Groups with more than one element that do not have any proper normal subgroups are referred to as *simple*.

DEFINITION 1.7 [Simple group]   *A group $G \neq 1$ is simple if it has no normal subgroups other than 1 and G.*

Example 1.5 [Center]
The *center* of a group G, often written as $Z(G)$, is defined as the set of those elements $z \in G$ that satisfy $gz = zg$ for all $g \in G$. It is easy to prove that $Z(G)$ is a normal subgroup of G for any group G.   ☐

Why are normal subgroups so interesting? Fixed a normal subgroup H of G, we may define an equivalence relation R in G as $g_1 R g_2$ *if and only if* $g_1^{-1}g_2 \in H$. The equivalence classes of this relation are called *H-cosets* and G/R is typically denoted by G/H and called *the quotient group* of G by H, referring to the fact that it also "inherits" the structure of G. For the order of G/H the suggestive notation $|G : H|$ is used and it is referred to as the *index* of H in G. If G is finite, we indeed have

$$|G| = |G : H| \cdot |H|.$$

Example 1.6 [Integers mod n: $\mathbb{Z}_n$]
Let n be any non-negative integer. The quotient of $(\mathbb{Z}, +)$ by the normal subgroup $n\mathbb{Z}$ is an Abelian group, which we will conveniently denote by $\mathbb{Z}_n$— rather than $\mathbb{Z}/n\mathbb{Z}$. Its elements (usually referred to as *classes modulo n*) can be represented by the set of remainders on division by n, i.e., $\{0, 1, \ldots, n-1\}$. We can define a product operation $\cdot$ on $\mathbb{Z}_n$ inherited from the product in $\mathbb{Z}$. It is easy to see that the set of units $\mathbb{Z}_n^*$ of $(\mathbb{Z}_n, \cdot)$ consists of exactly those elements $m \in \{0, 1, \ldots, n-1\}$ (respectively classes $[m]$) where m has no divisor in common with n. These numbers are said to be *relatively prime to n*.   ☐

Looking at this last example, we see that when dealing with $\mathbb{Z}_n$, it is always useful to have information about the numbers that divide n. Recall that a number $n > 1$ is *prime* if it has no positive divisors other than 1 and itself, and *composite* otherwise. Deciding whether a given integer is or is not prime, or finding prime divisors of a composite number are computational problems which play a central role for cryptographic constructions as relevant as the RSA encryption scheme (see Section 3.1.2).

In order to gain insight on an algebraic structure it is always useful to investigate the functions that preserve its specific properties.

DEFINITION 1.8 [Group homomorphism]   *Let $(G, \cdot)$ and $(H, \circ)$ be groups. Then a function $f : G \longrightarrow H$ is said to be a* group homomorphism *provided that $f(g \cdot h) = f(g) \circ f(h)$, for all $g, h \in G$.*

Group homomorphisms that are bijective are called *isomorphisms*; if there exists an isomorphism from a group G to a group H they are called *isomorphic*, which actually means they can be fully identified, as their underlying structure is identical. Every group of finite order can actually be proven to be isomorphic to a permutation group (this is called *Cayley's Theorem*).

Example 1.7 [Permutation groups]
Let X be a non-empty finite set. The set of all *permutations* on X, with the operation being composition, is a group usually referred to as the *symmetric group on X*. Common notations are $\mathrm{Sym}(X), S_X$ or, if X is of cardinality $n \in \mathbb{N}$, simply S_n. The order of S_n is $n!$ and each of its elements can be expressed as a product of *transpositions*, which are bijections that leave all elements in X fixed, except two of them which are swapped. That is, given two different selected elements $x_i, x_j \in X$, a transposition (i, j) is the bijection on X defined by $(i, j)x = x$ except for x_i, x_j and $(i, j)x_i = x_j$, $(i, j)x_j = x_i$. The subset of all permutations that can be written as the product of an even number of transpositions is a subgroup of S_X, called the *alternating* group on X and typically denoted by A_n, and its elements are often referred to as *even permutations*.   ▯

REMARK 1.1   To represent elements in a symmetric group S_n, *cycle notation* can be convenient, and we will use it: if $a_1, \ldots, a_r \in \{1, \ldots, n-1\}$ are pairwise different, then we will write $(a_1, \ldots, a_r)$ to represent the permutation that maps $a_1 \mapsto a_2$, $a_2 \mapsto a_3, \ldots, a_{r-1} \mapsto a_r$, $a_r \mapsto a_1$, and leaves all other elements pointwise fixed. Any permutation can be decomposed into a product of disjoint cycles.   ▯

Often, two different operations are defined on the same base set, as it happens with many structures we are most familiar with. An example is the case of a *ring*, that can be seen as an Abelian group equipped with an additional associative operation that is distributive over the Abelian group operation. Depending on the properties of the second operation (typically denoted as multiplication) rings are classified as commutative, with identity, division, etc.

DEFINITION 1.9 [Ring]   *A* ring *is a triplet* $(R, \cdot, +)$ *where* $(R, +)$ *is an Abelian group and* $(R, \cdot)$ *is a semigroup, such that* $\cdot$ *and* $+$ *fulfill the distributive law, i. e.,*

$$\forall a, b, c \in R : \ a \cdot (b + c) = a \cdot b + a \cdot c \ \text{ and } \ (b + c) \cdot a = b \cdot a + c \cdot a.$$

Rings may or may not have and identity element 1 for the $\cdot$ operation.

Example 1.8
The set of integers $\mathbb{Z}$ with the usual $+$ and $\cdot$ operations is a ring with identity 1. If we consider in $\mathbb{Z}_n$ the natural operations derived from $\mathbb{Z}$, we can prove $(\mathbb{Z}_n, +, \cdot)$ to be a ring with identity 1 as well. ▯

A class of groups arising in many different scenarios is that of *linear groups*. They are essential to the theory of group representations, arise in the study of symmetries of vector spaces, as well as in the study of polynomials.

Example 1.9 [Linear groups]
Let R be a commutative ring with identity element 1, $n \geq 1$ a positive integer, and denote by $\mathrm{GL}_n(R)$ the set of all $n \times n$ invertible matrices with entries in R. It is easy to see that $\mathrm{GL}_n(R)$ is a group, called the *general linear group over* R. Any group H that is isomorphic to a subgroup of a general linear group is called *linear*. Note that if V is a vector space of dimension n over a field F, then invertible linear transformations of V with standard composition constitute a group isomorphic to $\mathrm{GL}_n(F)$. ▯

A similar role to that played by normal subgroups in group theory is played by ideals within ring theory:

DEFINITION 1.10 [Ideal]   *Let R be a ring. A subset $I \subseteq R$ is a* left ideal *in R if given any natural number n, any n elements $r_1, \ldots, r_n$ from R, and any n elements $x_1, \ldots, x_n$ from I we have*

$$r_1 x_1 + \cdots + r_n x_n \in I.$$

Similarly, if for any such choice of elements we have

$$x_1 r_1 + \cdots + x_n r_n \in I$$

we say I is a right ideal. *If I is a left and right ideal we simply say it is an* ideal *in R, which is usually denoted by $I \trianglelefteq R$.*

Example 1.10

Let R be a ring with identity 1, and consider the structure naturally defined on the set

$$R[x] = \{a_0 + a_1 x + \cdots + a_n x^n \mid n \in \mathbb{N}_0, a_0, \ldots, a_n \in R\}$$

of polynomials of finite degree with coefficients from R. The set $R[x]$ is actually a ring with identity element 1. Fixing an element $r \in R$, define S as the set of polynomials from $R[x]$ which vanish on r. It is easy to see that S is an ideal in R, for if $p \in S$, then $p(r) = 0$, and clearly also $(q \cdot p)(r) = (p \cdot q)(r) = 0$ for any $q \in R[x]$. □

DEFINITION 1.11 [Field]   *A* field *is a commutative ring $(F, \cdot, +)$ with identity 1 where both $(F \setminus \{0\}, \cdot)$ and $(F, +)$ are Abelian groups.*

Field and ring homomorphisms can be defined as maps preserving the structure (as we did above for group homomorphisms), and in the same way isomorphisms are introduced as a basic tool to establish "algebraically identic" structures. Also, subrings and subfields may be defined in the same way as we did above for subgroups; they are simply derived from subsets preserving the corresponding algebraic structure.

Finite fields are of special interest within cryptography as being one of the major mathematical structures from which cryptographic constructions are derived.

THEOREM 1.1 [Existence of finite fields]

For every prime number p and every positive integer n there exists a field having exactly p^n elements. This field is usually denoted $\mathbb{F}_{p^n}$.

Furthermore, it can be shown that there are no other finite fields and that any two fields of order p^n are isomorphic. In particular, the structure $(\mathbb{Z}_p, +, \cdot)$ with p prime (see Example 1.6) gives us, up to isomorphism, all finite fields of prime order p.

1.2   Finite groups

In this section we review some important results and exemplify how a collection of finite groups is typically classified (following the exposition of [Hun89]). Most cryptographic constructions grounded in number theoretical problems are set on finite cyclic groups. Moreover, finite permutation and linear groups play an important role in many application scenarios; the essential results and terminology from this section will be useful to gain some insight on their structure.

Let us start with some facts related to the order of a finite group. Given a finite group G and a normal subgroup H hereof, we have already introduced the quotient group G/H and defined the *index* of H in G as its order $|G/H|$. If H is not normal, however, we define the *index* of H in G as the number of different left H-cosets, defined as sets $gH = \{gh \mid h \in H\}$, which can be proven to coincide with the number of right cosets $Hg = \{hg \mid h \in H\}$. The following result is thereafter easily derived.

THEOREM 1.2 [Lagrange's theorem]
Let G be a finite group and $H \leq G$. Then

$$|G| = |G : H| \cdot |H|.$$

In particular, if G is finite, the order of every element $g \in G$ divides the order of the group $|G|$.

As a consequence, we have

PROPOSITION 1.1
Let p be a prime number. A p-group is a group in which for each element $g \in G$, there exists an integer $k \geq 0$ so that the order of g is p^k. A finite group G is a p-group if and only if $|G|$ is a power of p.

The *Sylow theorems* help us depicting a bit better the structure of a finite group, scrutinizing the order of its elements and subgroups.

THEOREM 1.3 [First Sylow theorem]
Let G be a group of order $p^n m$, with $n \geq 1$, p prime and $\gcd(m, p) = 1$. Then G contains a subgroup of order p^i for each $1 \leq i \leq n$ and every subgroup of G of order p^k for $k < n$ is normal in some subgroup of order p^{k+1}.

Choosing $i = n$ in the first Sylow theorem, we obtain what is known as a *Sylow p-subgroup*.

DEFINITION 1.12   *Let G be a group and p a prime number. A subgroup P of G is said to be a* Sylow *p-subgroup if P is a maximal p-subgroup of G, namely, P is a p-group and if H is a p-group so that $P \leq H \leq G$ then it must be $H = P$.*

It is easy to see that if G is a finite group of order $p^n m$ with $n \geq 1$, p prime and $\gcd(m, p) = 1$, then H is a Sylow p-subgroup of G if and only if $|H| = p^n$. As a result, every conjugate of a Sylow p-subgroup is also a Sylow p-subgroup. If there is only one Sylow p-subgroup then it must be normal.

THEOREM 1.4 [Second Sylow theorem]
If H is a p-subgroup of a finite group G and P is any Sylow p-subgroup of G, then there exists $g \in G$ such that $H \subseteq gPg^{-1}$. In particular, any two Sylow p-subgroups of G are conjugate.

The third of the Sylow theorems gives information about the number of Sylow p-subgroups in a finite group.

THEOREM 1.5 [Third Sylow theorem]
If G is a finite group and p a prime, then the number of Sylow p-subgroups of G divides $|G|$ and is of the form $kp + 1$ for some non-negative integer $k \geq 0$.

When trying to decompose groups into smaller components, the concept of *direct product* is quite useful. For instance when looking at logarithmic signatures as discussed in Section 4.6, one can easily obtain a logarithmic signature for a finite direct product of finite groups from logarithmic signatures of the constituting factors.

DEFINITION 1.13 [External direct product]   *Let $G_1, \ldots, G_n$ be groups and $n \in \mathbb{N}$ a positive integer. We define their* external direct product *to be the cartesian product set $G_1 \times \cdots \times G_n$ equipped with component-by-component multiplication of n-tuples defined in the natural way.*

It is easy to see that the external direct product of groups is itself a group. Furthermore, for each i we have a subgroup of $G_1 \times \cdots \times G_n$,

$$\hat{G}_i = (\epsilon_1) \times (\epsilon_2) \times \cdots \times G_i \times \cdots (\epsilon_n)$$

which is isomorphic to G_i. This simple observation allows us to formulate the following result.

THEOREM 1.6 [Internal direct product]
Let G be a group and $H_1, \ldots, H_n$ subgroups fulfilling

- *each H_i is normal in G,*

- *the product set $H_1 \cdots H_n$ is equal to the set G, and*

- *each $g \in G$ decomposes uniquely as $g = h_1 \cdots h_n$, with $h_i \in H_i$ for $i = 1, \ldots, n$.*

Then G is isomorphic to the direct product group $H_1 \times \cdots \times H_n$.

Using the above theorems and other simple techniques it is easy to classify (up to isomorphism) all groups of order pq for p, q primes and all groups of small order $n \leq 15$. We start by looking at the Abelian case.

THEOREM 1.7 [Fundamental theorem for finite Abelian groups]
Let G be a finite (non-trivial) Abelian group. Then G can be written as the direct product of cyclic groups of orders $m_1, \ldots, m_t$, with $m_1 > 1$ and m_i dividing m_{i+1} for $i = 1, \ldots, t - 1$.

For an Abelian group whose order is a product of two different primes, Theorem 1.7 immediately implies that this group must be cyclic. In fact, more can be shown.

PROPOSITION 1.2
Let p and q be primes with $p > q$. If q does not divide $p-1$, then every group of order pq is isomorphic to a cyclic group of pq elements. Otherwise, there are exactly two distinct groups or order pq, the cyclic group of pq elements and a non-Abelian group K generated by elements c and d such that $|c| = p$, $|d| = q$ and $dc = c^s d$, where $s \not\equiv 1 \pmod{p}$ and $s^q \equiv 1 \pmod{p}$.

Typically, if (as happens above) a group G is fully defined by generating elements $x_1, \ldots, x_t$ subject to equations $e_1, \ldots, e_r$, we may represent it by

$$\langle x_1, \ldots, x_n \mid e_1, \ldots e_r \rangle .$$

We will have a chance to study such *finitely presented* groups further in Section 6.1. Various proposals that have been made over the last years to use group theory in cryptography work explicitly with this type of group representation.

From Proposition 1.2 it follows that if p is an odd prime, then every group of order $2 \cdot p$ is isomorphic either to the cyclic group $\mathbb{Z}_{2p}$ or to the *dihedral group D_p*, defined by:

$$D_p = \langle a, b \mid a^2 = 1, b^p = 1, ab = b^{-1}a \rangle .$$

PROPOSITION 1.3

There are, up to isomorphism, exactly two distinct non-Abelian groups of order 8: the dihedral group D_4 and the quaternion group Q_8 defined by

$$Q_8 = \langle a, b \mid a^4 = 1, a^2 = b^2, b^{-1}ab = a^{-1} \rangle.$$

With the above results we are almost able to classify all finite groups of order ≤ 15 up to isomorphism. A key piece missing is to know the possible structure of groups of order $12 = 2^2 \cdot 3$ that can occur. One can show the following.

PROPOSITION 1.4

Up to isomorphism, there are exactly three distinct non-Abelian groups of order 12:

- *the dihedral group D_6,*

- *the alternating group A_4, and*

- *the group*
$$T = \langle a, b \mid |a| = 6, b^2 = a^3 \text{ and } ba = a^{-1}b \rangle.$$

A natural question to ask is to classify *all* finite groups up to isomorphism. Even after reducing the question to classifying all simple groups, this turns out to be a highly non-trivial problem. Tackling this question requires a major research endeavor, and has involved many years of research before the *Classification Theorem* was in place. A short paper by Aschbacher [Asc04] gives an idea of the difficulty of establishing and proving the Classification, which according to [Asc04] *is the most important result in finite group theory.*

1.3 Summary and further reading

In this chapter we have given some basic definitions and results that are foundational for this book. For a good understanding of the upcoming parts some mathematical maturity will still be needed, and we encourage the reader to dive into the following sections and catch up with mathematical foundations along the way. Having a cryptographic application in mind may sometimes provide additional motivation to strengthen one's working knowledge in a mathematical subject.

There are uncountable[1] great texts on group theory, finite fields, and number theory which provide an excellent source for building a solid basis toward

---

[1]Figuratively speaking.

cryptographic applications. To name a few, we mention here Rotman's introductory text on group theory [Rot95] and some "classic" texts on computational aspects of group theory [Sim94, ECH⁺92]. Regarding finite fields, a canonical reference is a book by Lidl and Niederreiter [LN97] and if looking for application-geared texts, books by Shparlinski [Shp92] or Menezes et al. [MBG⁺93] can be recommended highly.

Following a "minimalist" approach in this chapter, we have hardly glimpsed at number theory in our exposition, and a great text to get into the subject is [Shp99], whereas the basics on how to compute in $\mathbb{Z}_n$ can be found in any good book on discrete mathematics, such as [Ros02].

1.4 Exercises

Exercise 1 *Suppose an element u of a monoid M has a right inverse x, i. e., we have $ux = \epsilon$, and a left inverse y, i. e., $yu = \epsilon$. Prove that in this situation the element u is actually a unit.*

Exercise 2 *Let G be a group. For each $g \in G$, consider the map*

$$\begin{aligned} f_g : G &\longrightarrow G \\ h &\longmapsto g^{-1}hg \end{aligned}.$$

(a) *Prove that f_g is an automorphism of G, i. e., an isomorphism from G into itself.—The map f_g is commonly called the* inner automorphism *induced by g.*

(b) *Now define*

$$\mathrm{Inn}(G) = \{f_g \mid g \in G\},$$

and show that $\mathrm{Inn}(G)$ is a subgroup of $\mathrm{Sym}(G)$.

(c) *If we denote by $\mathrm{Aut}(G)$ the group of all automorphisms of G, prove that $\mathrm{Inn}(G) \trianglelefteq \mathrm{Aut}(G)$.*

Exercise 3 *Which subgroup of $(\mathbb{Z}, +)$ is generated by 2 and 6? And the one generated by p and $p + 2$ if p is a prime number?*

Exercise 4 *Prove that if G is a cyclic group of prime order, then each $g \in G$ different from the identity 1 is a generator of G.*

Exercise 5 *Let n be any positive integer. Two integers z, u are said to be congruent modulo n (denoted $z \equiv u \pmod{n}$) if they have the same remainder on division by n. Prove that this notion defines an equivalence relation on $\mathbb{Z}$. Furthermore, see that its corresponding quotient is the set described in Example 1.6.*

Exercise 6 *Prove that if H is a normal subgroup of order p^k of a finite group G, then H is contained in every Sylow p-subgroup of G.*

Exercise 7 *Euler's totient function φ assigns to a positive integer n the number of positive integers less than n that have no common divisors with n. As a result, the group of (multiplicative) units in $\mathbb{Z}_n$ has order $\varphi(n)$.*

Show that φ is multiplicative in the sense that $\varphi(mn) = \varphi(m) \cdot \varphi(n)$ provided that m and n are relatively prime. Conclude that if $n = p_1^{n_1} \cdots p_k^{n_k}$ with pairwise distinct prime numbers p_i, then

$$\varphi(n) = p_1^{n_1} \cdot \left(1 - \frac{1}{p_1}\right) \cdot \ldots \cdot p_k^{n_k} \cdot \left(1 - \frac{1}{p_k}\right).$$

Exercise 8 *Let $n > 1$ be a positive integer and choose $1 \le a < n$ such that n is relatively prime to a. Using the definitions from the previous exercises, prove that $a^{\varphi(n)} \equiv 1 \pmod{n}$—this statement is also known as* Euler's theorem.

Exercise 9 *Prove that for any $m \in \mathbb{N}$ the set $m\mathbb{Z} = \{mz \mid z \in \mathbb{Z}\}$ is an ideal in the ring of integers. An ideal I of a commutative ring R is called a* principal ideal *if there exists an element $a \in I$ such that $I = \{a \cdot b \mid b \in R\}$. Prove that every ideal of $\mathbb{Z}$ is principal.*

Exercise 10 *Show that the dihedral group D_4 is not isomorphic to the quaternion group Q_8.*

Chapter 2

Basics on complexity

2.1 Complexity classes

Complexity theory is concerned with the intrinsic hardness of computational tasks. Measuring such hardness accurately is a crucial step toward a secure cryptographic design. Traditionally, algorithmic problems are formalized as *decision* or *search* problems, depending on the output value that is considered to constitute a solution of a concrete instance of the problem. If this output is just one bit, we speak of a *decision* problem, whereas for *search* problems the output belongs to a set of larger cardinality—which in particular allows for capturing the situation in which a solution does not exist. Further, most problems in cryptology actually belong to the class of *promise problems*, which is a generalization of decision/search problems where the input is promised to belong to a specific subset of the set of all possible inputs (see [Gol08, Chapter 2]). We will phrase the definitions of complexity classes in terms of decision problems; they can however be adapted for search and promise problems.

Example 2.1
Let us state the presumed input/output of an algorithm solving some elementary stated (yet intricate) problems related to number theory.

Primality Testing (decision):
    Input: $s \in \{0,1\}^+$, output: $b \in \{0,1\}$, where

$$b = \begin{cases} 1 \text{, if the integer whose binary representation is } s \text{ is prime} \\ 0 \text{, otherwise.} \end{cases}$$

Factoring (search):
    Input: $s \in \{0,1\}^+$, output: $d \in \{0,1\}^+$, where d is the binary expansion of a proper divisor of the integer whose binary representation is s, if there is any, and 1 otherwise.

RSA-factoring (promise search):
    Input $s \in \{0,1\}^+$, output: $d \in \{0,1\}^+$, where d is the binary expansion

of a prime factor of the integer whose binary representation is s, provided that this input value is the product of two primes.

□

The standard strategy to evaluate the complexity of a computational task is to analyze a *worst-case* scenario, that is, to classify problems and algorithms considering the worst possible instance in terms of computing "time." In this case, "time" reflects the number of basic bit operations, though often the only computations taken into account are multiplications (additions/subtractions and modular reductions are disregarded, and exponentiations are computed in terms of products—see Exercise 11). The choice about the elementary operations that are counted is important for the practical value of complexity statements—it is desirable that the model choices made here reflect the cost in an actual implementation.

Another approach that is often (more or less explicitly) used when aiming at a cryptographic design is *average-case complexity*. As O. Goldreich points out in [Gol08], this paradigm may actually be better referred to as *typical-case complexity*, for its spirit is to analyze the time consumption ignoring instances that may be considered exceptional—not only rare, but also irrelevant in some (often practical) sense. As much as this approach is of interest to this book, we restrict our study here to the more classical *worst-case* scenario.

DEFINITION 2.1 [The class $\mathcal{P}$]   *A decision problem P is in the class $\mathcal{P}$ of polynomial time problems, if there exists a polynomial $p(n)$ and an algorithm $\mathcal{A}$ such that if an instance of P has input length bounded by n, $\mathcal{A}$ outputs a correct answer in time bounded by $p(n)$. This holds for any natural number n encoding the input length for P.*

The above definition can also be phrased saying that there exists an algorithm $\mathcal{A}$ for the problem P which is polynomial time (namely, the number of bit operations it performs on an input of length n is bounded by $p(n)$ for some polynomial p). The primality testing problem from Example 2.1 is known to be in $\mathcal{P}$, whereas factoring and RSA-factoring are not known to be in this class (and often presumed not to be).

The class $\mathcal{P}$ comprises computational problems that may "in principle" be solved rapidly in practice. Notwithstanding this, one has to refine the analysis before making risky assertions in this sense, as the practical implications of the fact that a problem lies in $\mathcal{P}$ are sometimes limited. Indeed, it can well be the case that the best algorithm for a problem in $\mathcal{P}$ is terribly slow in practice—the degree and leading coefficient of the polynomial p from the above definition may be a good hint for identifying such cases.

Usually, the algorithm $\mathcal{A}$ from the above definition is assumed to be deterministic. If we generalize the above definition and allow $\mathcal{A}$ random selections and then impose that "yes" answers are always right and "no" answers are correct with probability greater than $1/2$, we speak of the *probabilistic polynomial time* (class $\mathcal{RP}$). Further, the class $\mathcal{BPP}$ is informally the one capturing the case when $\mathcal{A}$ is randomized and both the "yes" and "no" answers are correct with probability greater than $1/2$.

Another twist of the above notion which is particularly relevant in cryptography, is the case for which a *different* algorithm solves the problem depending on the instance input size.

DEFINITION 2.2 [Non-uniform polynomial time]   *A decision problem P is solvable in* non-uniform polynomial time *(denoted $P \in \mathcal{P}\backslash Poly$) if there exists a polynomial $p(n)$ and a sequence of algorithms $\{\mathcal{A}_n\}_{n \in I}$, for some index set $I \subseteq \mathbb{N}$ such that if an instance of P has input length bounded by $n \in I$, $\mathcal{A}_n$ outputs a correct answer in time bounded by $p(n)$.*

It can be proved that both the class $\mathcal{P}$ and the class $\mathcal{BPP}$ are contained in $\mathcal{P}\backslash Poly$.

DEFINITION 2.3 [The class $\mathcal{NP}$]   *A decision problem P is in the class $\mathcal{NP}$ if given any instance of P an evidence establishing the "yes" answer can be checked with a polynomial time algorithm.*

Clearly, we have $\mathcal{P} \subseteq \mathcal{NP}$, but one of the most bewildering open problems in theoretical computer science is whether both classes are actually equal or not.

Another important concept is that of polynomial reduction. Informally, we say a problem P_1 reduces to a problem P_2 in polynomial time provided that there exists an algorithm for P_1 which uses at most polynomially many calls to a P_2-*oracle*. At this, an *oracle* is an algorithm that given the input of a P_2 instance provides its solution (without revealing anything else, i.e., in a black-box manner). Typically, the time taken by the oracle is not taken into account when evaluating the resources of the algorithm for solving P_1. An important issue is whether the provided reduction is or is not *tight*. The informal idea is that a tight reduction states that P_1 and P_2 have essentially the same complexity, and thus a solution for P_2 is effective to solve P_1. If the reduction is not tight it may happen that the algorithm we construct for P_1 using P_2 as a black box is atrociously slow; as a result an efficient algorithm for P_2 may not be too helpful in practice to solve P_1. Using the notion of polynomial reduction, we may define another relevant class.

DEFINITION 2.4 [$\mathcal{NP}$-complete problems]   *A decision problem P in $\mathcal{NP}$ is said to be $\mathcal{NP}$-complete if every other problem in $\mathcal{NP}$ can be reduced to P in polynomial time.*

Even though complexity arguments are at the core of most cryptographic proofs of security, some care must be taken before trusting the security of a scheme based on a *theoretically* (say $\mathcal{NP}$-complete) hard problem. A popular example of what may go wrong is what happened with the Merkle–Hellman encryption scheme from [MH78]. This construction is based on a combinatorial problem called *the subset sum problem*, known to be $\mathcal{NP}$-complete. However, as A. Shamir evidenced in [Sha84], the concrete instances of the subset sum problem relevant to Merkel–Hellman encryption can be solved with a polynomial time algorithm.

2.2 Asymptotic notation and examples

When comparing the running time of algorithms, it is often convenient to disregard parts of the expressions that make explicit the concrete number of bit operations performed, for they become irrelevant as the input size of the algorithms grows. Hence, as complexity statements are asymptotic in nature, it is customary to adopt a special notation that facilitates comparisons between different running times.

DEFINITION 2.5 [Big-$\mathcal{O}$ notation]
Consider two functions $f : \mathbb{N}_0 \longrightarrow \mathbb{R}_{\geq 0}$ and $g : \mathbb{N}_0 \longrightarrow \mathbb{R}_{\geq 0}$ that map non-negative integers to non-negative real numbers. If for some $n_0 \in \mathbb{N}_0$ and for some positive constant C we have

$$f(n) \leq C \cdot g(n), \text{ for all } n \geq n_0,$$

then we write $f = \mathcal{O}(g)$.

This notation is useful when we want to simply disregard parts of our "operation counting" without losing relevant information, e. g., writing $\mathcal{O}(n^3)$ instead of $4n^3 + 12n^2 - 15n + 3$ can be clever. At the same time, using $\mathcal{O}(n^3)$ instead of e^{-n} seems, despite being technically correct, not particularly helpful.

Example 2.2
Note that the length of the bit representation of a natural number m is in $\mathcal{O}(\ln m)$. Then

- we say multiplying natural numbers m and n is "quadratic," as there is an algorithm that receives as input binary representations of m and n and then will require $\mathcal{O}(\ln m \cdot \ln n)$ operations to compute the product $m \cdot n$.

- dividing a natural number a by a natural number b (computing quotient and remainder, with $0 < b < a$) is also quadratic. If a has a binary representation of length ℓ, we can construct an algorithm taking $\mathcal{O}(\ell^2)$ operations for this task.

$\square$

Example 2.3 [(Extended) Euclidean algorithm]
Given two positive integers $0 < a < b$, computing their greatest common divisor

$$d = \gcd(a, b)$$

along with two integers α and β satisfying *Bézout's identity*

$$\alpha \cdot a + \beta \cdot b = d \tag{2.1}$$

can be solved in time $\mathcal{O}(\ln a \cdot \ln b)$. A famous algorithm for doing so is the *(extended) Euclidean algorithm*. It is based on the fact that, if the division equation states that

$$a = bq + r \text{ with } 0 \le r < b,$$

then $\gcd(a, b) = \gcd(b, r)$. Now, the idea is to perform successive divisions until we have an exact result:

$$
\begin{array}{llll}
a & = & q_0 b + r_1 & 0 < r_1 < b \\
b & = & q_1 r_1 + r_2 & 0 < r_2 < r_1 \\
r_1 & = & q_2 r_2 + r_3 & 0 < r_3 < r_2 \\
& \vdots & & \vdots \\
r_{j-1} & = & q_j r_j + r_{j+1} & 0 < r_{j+1} < r_j \\
& \vdots & & \vdots \\
r_{l-2} & = & q_{l-1} r_{l-1} + r_l & 0 < r_1 < r_{l-1} \\
r_{l-1} & = & q_l r_l + r_{l+1} & 0 < r_{l+1} < r_l \\
r_l & = & q_{l+1} r_{l+1}
\end{array}
$$

Here, the greatest common divisor of a and b equals the last non-zero remainder, i.e., r_{l+1}. To find a linear expression of the form (2.1), we can backtrack along the above equations, bottom to top, getting linear expressions of the form $\gcd(x, y) = c_x x + c_y y$, culminating with the integers α, β we were looking for. $\square$

Example 2.4 [Discrete logarithm problem]
The *discrete logarithmic problem* may be stated as the following search problem. On input a positive integer n, a generator g of a cyclic group of order n, and a uniformly at random chosen element h in the group $G = \langle g \rangle$, find the *exponent* $\alpha \in \{0, \ldots, n-1\}$ such that $g^\alpha = h$. Depending on the concrete group

at hand and on how it is represented, the complexity of the best available algorithms for this problem varies significantly.

- Take as input a generator g of the additive group $\mathbb{Z}_n$. Then, solving the discrete logarithm problem is equivalent to solving the equation

$$gx \equiv h \pmod{n},$$

which can be done in polynomial time (see Exercise 15).

- If the input is a generator of a cyclic group that can be seen as a subgroup of $(\mathbb{F}_q, \cdot)$, for $\mathbb{F}_q$ a finite field, then no polynomial time algorithm for this problem is known (generally speaking the best we can do so far is *subexponential*, see [AD94]).

- The discrete logarithm problem in the general linear group over the finite field of q elements $\mathrm{GL}_n(\mathbb{F}_q)$ can be reduced to the problem of finding solutions in several small extension fields $\mathbb{F}_{q^{m_j}}$, where $m_j \leq n$, $j = 1, \ldots, s$, and s is the number of irreducible factors in $\mathbb{F}_q[x]$ of the characteristic polynomial of the generating matrix g (see [MW97]). As a cryptanalytic insight, solving discrete logarithms in this group is of comparable difficulty as doing it in $\mathbb{F}_{q^n}$.

$\Box$

2.3 Summary and further reading

In this chapter we have reviewed the basics of complexity theory, defining the main complexity classes we will be confronted with and discussing a few examples to get used to asymptotic notation. Again, our approach has been rather minimalist and those willing to know more on these fascinating topics are encouraged to take a look in some of the texts we suggest below. We also recommend the interested reader to take a closer look at the problems we mentioned in Example 2.1. A survey on factoring problems can be found in [Len11], and [AKS04] offers a beautiful proof evidencing primality testing to be in $\mathcal{P}$.

Goldreich's book [Gol08] gives a thorough and up-to-date presentation on the theory of complexity, whereas [Gol99] focuses on (basic and advanced) topics encompassing *computational pseudorandomness*, *probabilistic proof systems*, and cryptography. Another book, which for the beginner may perhaps be easier to read, has been written by Shoup [Sho06]. For readers interested in algorithmic problems arising from group theory, the handbooks [Sim94, ECH$^+$92, HEO05] can be recommended.

In our short exposition above we have only considered a *classical* computational model (built under the assumption that computations are performed via Turing machines with access to a random tape). Thus, all *algorithms* mentioned in this chapter are assumed to run on a standard (classical) computer. There are, however, different models that take into account non-standard computing devices, such as DNA computers or quantum computers. Quantum complexity and quantum information theory are nowadays flourishing research subjects, and the practical implications of these theories, especially in the field of cryptography, are already noticeable. The handbook [NC00] is an excellent introduction to the main ideas and techniques of quantum computation and information.

2.4   Exercises

Exercise 11 *Let $1 < a < n$ and e be positive integers and ℓ the bitlength of n. If the binary expansion of e is*

$$e = e_0 + e_1 \cdot 2 + e_2 \cdot 2^2 + \cdots + e_{k-1} \cdot 2^{k-1},$$

then one can compute $a^e \pmod{n}$ in time $\mathcal{O}(\ell^2 k)$ by means of a procedure known as square and multiply. *It exploits the fact that*

$$a^e \equiv \prod_{e_j=1} (a^{2^j} \pmod{n}) \pmod{n}.$$

Try to design (and implement) an algorithm for this procedure and confirm the $\mathcal{O}(\ell^2 k)$ bound.

Exercise 12 *Prove that we could alternatively have defined the class $\mathcal{P}$ as the class of problems P for which there exists a positive constant $c > 0$ and an algorithm $\mathcal{A}$ for solving P in time $\mathcal{O}(n^c)$.*

Exercise 13 *Check the claimed complexity for the Euclidean algorithm from Example 2.3.*

Exercise 14 *For any integer a and any odd prime p the* Legrende symbol $\left(\frac{a}{p}\right)$ *is defined to be*

$$\left(\frac{a}{p}\right) = \begin{cases} 0, & \text{if } a \equiv 0 \pmod{p} \\ 1, & \text{if } a \not\equiv 0 \mod p \text{ and for some integer } x, \ a \equiv x^2 \pmod{p} \\ -1, & \text{otherwise.} \end{cases}$$

Essentially, the Legendre symbol tells us if a is a quadratic residue *modulo the prime number p. Now, for any integer a and any positive odd integer* $m = p_1^{\alpha_1} p_2^{\alpha_2} \cdots p_k^{\alpha_k}$ *the Jacobi symbol* $\left(\frac{a}{m}\right)$ *is defined as the product of the Legrende symbols corresponding to the pairwise distinct prime factors of m, namely*

$$\left(\frac{a}{m}\right) = \left(\frac{a}{p_1}\right)^{\alpha_1} \left(\frac{a}{p_2}\right)^{\alpha_2} \cdots \left(\frac{a}{p_k}\right)^{\alpha_k}.$$

In [CP05, Algorithm 2.3.5.], the following algorithm is given for computing the Jacobi symbol $\left(\frac{a}{m}\right)$ *where m is an odd integer and a is an integer with* $|a| < m$. *Analyze its worst-case complexity.*

1. # Reduction mod m

    $a = a \pmod{m}$;

    $t = 1$;

    while $(a \neq 0)$ {

        while $(a$ even$)$ {

            $a = a/2$;

            if $m \pmod 8 \in \{3, 5\}$

                then $t = -t$;

        }

        $(a, m) = (m, a)$; # swap variables

        if $a \equiv m \equiv 3 \pmod 4$

            then $t = -t$;

        $a = a \pmod{m}$;

        }

2. # Termination

    if $m = 1$

        then return t

        else return 0.

Exercise 15 *Use the extended Euclidean algorithm from Example 2.3 to prove that finding the multiplicative inverse of a unit a in* $\mathbb{Z}_n$ *is quadratic.*

Exercise 16 *Let* $f_1(x), \ldots, f_n(x)$ *be univariate polynomials over a finite field* $\mathbb{F}_q$. *Sketch an algorithm for testing membership of elements from* $\mathbb{F}_q[x]$ *in the ideal*

$$\{g \mid g = g_1 f_1 + \cdots g_n f_n, \text{ where } g_i \in \mathbb{F}_q[x] \text{ for } i = 1, \ldots, n\}$$

they generate.

Chapter 3

Cryptology: An introduction

3.1 A short historical overview

Cryptology is the (art and) science devoted to information protection in a wide sense; it provides solutions for storing, sharing, communicating, and even computing with data when certain security guarantees are required. It has both a constructive and a destructive face. In the classical terminology, *cryptography* deals with the design of secure schemes, while *cryptanalysis* audits the claimed robustness of cryptographic designs. Nowadays many people do not make a strict separation between cryptography and cryptanalysis, and cryptography is often understood as a synonym for cryptology.

One could say cryptology is to some extent inherent to language, and it is hard to give a precise date of when it all started. What is historically clear, however, is that different events and technological developments had a strong influence on the way people communicated and had as a result striking consequences on the way cryptology was used and understood. Below we will briefly go through some of these milestones, describing informally a few basic cryptographic schemes. We will see a significantly more rigorous approach to discussing cryptographic schemes in the next part of the book.

3.1.1 Historical encryption schemes

Allowing for private communication was the main (if not the only) cryptographic goal until the second half of the twentieth century. Up until that point, cryptography was devoted to the design of encryption schemes—or *ciphers*— that would allow a sender to construct dedicated messages which, hopefully, could only be read by legitimate receivers. Sender and receiver would for that purpose hold some special piece of information (a key) or token (encryption/decryption device). As a result, encryption and decryption processes were *symmetric*, in the sense that both sender and receiver performed basically the same operations, only that the recipient might (un)do the sender's process in reverse order. Similarly, sender and receiver held identical secret tokens for enabling secure communication. This symmetry between sender and receiver is the main idea underlying *secret-key* cryptography, which is also commonly referred to as *symmetric* cryptography: all involved users hold identical information/resources.

A strategy of the early days was to assume that the encryption method was unknown to a potential adversary, but it was soon realized that this assumption should not be vital for security. Rather than having to rely on the assumption that an adversary cannot capture a device that is used for encryption or decryption, security should be based on the fact that a shared secret key is unknown to the adversary. This idea (together with some other statements towards good cryptographic practice), was made concrete by Auguste Kerckhoffs in the nineteenth century in a principle that has been named after him [Ker83]:

> *Il faut qu'il n'exige pas le secret, et qu'il puisse sans inconvénient tomber entre les mains de l'ennemi;*

An approximate translation into English offered by Fabien Petitcolas [Nic14] is that *[t]he system must not require secrecy and can be stolen by the enemy without causing trouble.* Encryption transformations used until the twentieth century essentially combined two types of strategies:

- **Substitution:** Each letter of the plaintext alphabet is substituted with a letter from the ciphertext alphabet via a secret rule (the *key*). Note that the two alphabets may very well coincide. The simplest way to do so is the so-called *shift cipher*, where each letter of the plaintext alphabet is replaced by the one located k places on in the same alphabet (wrapping around at the end of the alphabet in the obvious way), k being the secret key. Julius Caesar was known to use this method with $k = 3$. Another example is the *Vigenère cipher* from the sixteenth century, which applies several different substitution rules at the same time, thus being called a *polyalphabetic substitution cipher*.

- **Permutation:** The plaintext words are parsed into subwords—say, of length $t \in \mathbb{N}$, whose characters are permuted using a secret permutation $\sigma \in S_t$. A simple example in which the complete text is permuted is the spartan *scytale* (500 B.C.): it is formed by a ribbon that is wrapped around a dowel of a particular (secret) diameter and length. The plaintext is written on the ribbon wrapped on the dowel, so that, hopefully, a dowel of the same diameter (which is actually the key) is needed to read off the message.

It is clear that such basic strategies, would not hide the language statistics sufficiently, and as a result all these historical methods become insecure when encrypting natural language and sufficient amounts of ciphertext are available. Notwithstanding this, until today the design of symmetric encryption schemes makes extensive use of (more or less sophisticated) substitution and permutation techniques.

Substitution ciphers were mechanized, achieving remarkable robustness. A famous example is the *Enigma*, a rotor machine—or more accurately a family of rotor machines—invented in the 1920s which played a central role in World War II. Several different variants of this machine were used, but basically Enigma implemented different substitution ciphers using moving rotors, which resulted in a 26-alphabet polyalphabetic substitution cipher per rotor. We will not go into the specifics of the fascinating history of Enigma and its cryptanalysis here, but let us at least take a brief look at how rotors are combined to implement a polyalphabetic substitution.

Example 3.1

Given five rotors indexed with alphabets in the order

> EKMLFGDQVZNTOWYHXUSPAIBRCJ
> AJDKSIRUXBLHWTMCQGZNPYFVOE
> BDFHJLCPRTXVZNYEIWGAKMUSQO
> ESOVPZJAYQUIRHXLNFTGKDCMWB
> VZBRGITYUPSDNHLXAWMJQOFECK

a machine like the Enigma is, say every day, initialized with three of them in a certain position. For instance we may encode the following initial position as CRE:

> CQGZNPYFVOEAJDKSIRUXBLHWTM
> RTXVZNYEIWGAKMUSQOBDFHJLCP
> ESOVPZJAYQUIRHXLNFTGKDCMWB

Then, if we encrypt the letter A it will first go through the first rotor and encode itself as C, then it will go to X from the second rotor and finally it will come out encrypted as M. The Enigma involved a special type of rotor, a *reflector*, which implemented a fixed-point-free permutation and was used to send the "intermediate ciphertext" a second time through the preceding rotors—in reverse order. Each time the first rotor acted, its initial position changed (according to a ring hitting a notch), and this would have an effect on the position of the second rotor, which likewise influenced the position of the third rotor. For the Enigma machine, a plugboard offered a further refinement. It induced a swap of two letters on each encryption/decryption. All these settings translate into a key space that even by today's standards is rather substantial, and cryptanalyzing the Enigma was a highly non-trivial achievement. □

During World War I, an encryption method patented by Gilbert Vernam was extensively used—the *one-time pad*. Its structure is very simple. The plaintext message is encoded as a sequence over the binary field $\mathbb{F}_2$ (a bit-string), and encryption is performed bit by bit, adding each bit of the plaintext with a corresponding bit of the key (modulo 2). It is essential that keys are one-use and key bits are generated independently and uniformly at random for each encryption. The experienced robustness of the one-time pad was theoretically justified by Claude Shannon, who published a foundational paper in the Bell Labs Technical Journal in 1949 [Sha49] introducing the term *perfect secrecy*, which captures the property of an encryption procedure for which ciphertexts do not reveal any information about the corresponding plaintext—apart from the length of the plaintext. Shannon's theory is central to modern cryptology and his visionary work is the seed of modern *provable security* (cf. Chapter 4).

It is not hard to see that the one-time pad meets this requirement, as given a truly random key which is used only once, a ciphertext can be translated into any plaintext of the same length, and all are equally likely. Perfect secrecy is obviously a very attractive security guarantee. It has, however, a crucial drawback: the key space must at least be as large as the message space (which, in particular, if keys and messages are of fixed size implies that the key must be at least as long as the message). This limitation can be proved to be inherent to perfect secrecy (see [KL07, Chapter 2]).

3.1.2 Public-key cryptography

All the encryption methods we have mentioned so far had two things in common.

- They were constructed primarily for military use.

- They rely on a secure pre-distribution of secret keying material.

In 1976 two researchers, Whitfield Diffie and Martin Hellman, published a paper [DH76] depicting a method for public key agreement—two entities could, via a public communication channel, establish a secret common key for securing their subsequent communication. Their proposal is described informally in Figure 3.1 where, as it will happen in the rest of the book, $s \leftarrow S$ stands for *s is selected uniformly at random from S*. The set S will always be finite when we use this notation. Of course, for this protocol to offer any meaningful security guarantees, we will in particular need the discrete logarithm problem in the group G to be hard.

Once Alice and Bob share a common key $K_{AB} \in G$ in the finite group G, they may exchange securely a group element $m \in G$ masked as $K_{AB} \cdot m$, i.e., in

Let G be a public cyclic group of prime order q, with g a public generator of G.

1. Alice chooses $a \leftarrow \mathbb{Z}_q$ and computes $h_1 = g^a$

2. Alice sends h_1 to Bob

3. Bob chooses $b \leftarrow \mathbb{Z}_q$ and computes $h_2 = g^b$

4. Bob sends h_2 to Alice

5. Alice and Bob both compute the group element

$$K_{AB} = g^{ab} = h_1^b = h_2^a$$

FIGURE 3.1: Diffie–Hellman key agreement.

a *one-time pad fashion.*[1] In fact, a general encryption procedure is naturally derived from the Diffie–Hellman key agreement. Just perform all steps of the key agreement and interpret the first two steps as Alice transmitting her *public key* to Bob. Then, in Step 4 of Figure 3.1, let Bob send not only h_2 but also $K_{AB} \cdot m$. In other words, the pair

$$(h_2, K_{AB} \cdot m) \in G \times G$$

is the encryption of m under Alice's public key h_1. This method is the El-Gamal encryption scheme proposed in [ElG85], which we will get back to in Section 4.2. Prior to ElGamal encryption, another public-key asymmetric encryption method was proposed by Ronald Rivest, Adi Shamir, and Leonard Adleman: RSA encryption [RSA78]. We describe it informally in Figure 3.2.

The robustness of this so-called "textbook RSA" relies on the hardness of inverting a given RSA *function* defined by

$$RSA_{N,e}(x) = x^e \pmod{N}.$$

This problem is strongly related to the problem of factoring the RSA module N, but it is not at all obvious if the two problems are equivalent. Even though it is fair to say that RSA is the most extensively deployed encryption scheme, one should also realize that nowadays versions of RSA look quite different from the scheme described in the late 1970s—and there are good reasons for that, as we will see in Chapter 4. For the special case when $e = 2$ the corresponding scheme (known as Rabin encryption [Rab79]) has a nice security reduction. Namely, the problem of decrypting Rabin ciphertexts from

---

[1] The careful reader may notice that we actually want more than K_{AB} being unknown to the adversary—the key K_{AB} should "look like" a uniformly at random chosen group element. We will look into this more carefully when discussing provable guarantees.

Alice computes:

- N, a product of two primes p, q, with $\log_2 p \approx \log_2 q$,

- e, a unit from $\mathbb{Z}_{\varphi(n)}$.

The pair (N, e) is the public key, while e's multiplicative inverse $d = e^{-1} \in \mathbb{Z}^*_{\varphi(n)}$ is the secret decryption key.

1. Bob, the sender, encrypts $m \in \mathbb{Z}_N$ as $c = m^e \pmod{N}$.

2. Alice, holding d, decrypts by computing

$$c^d = m^{ed} = m \text{ in } \mathbb{Z}_N.$$

FIGURE 3.2:   RSA encryption.

the public information reduces polynomially to the problem of factoring the public modulus N, and this reduction is tight.

THEOREM 3.1 [Security of Rabin encryption]
Let $N \in \mathbb{Z}$ be the product of two odd primes $p \neq q$. Let $\mathcal{A}$ be an algorithm for finding one of the solutions of the equation

$$y^2 \equiv c \pmod{N}$$

in time $T(N)$. Then, there exists an algorithm for factoring N requiring $2T(N) + 2\log_2 N$ steps.

The interested reader can find a proof of the above theorem in the original paper [Rab79]. An outline of the proof is as follows: there are exactly four solutions to the congruence equation $x^2 \equiv c \pmod{N}$ which we may denote $k, -k, \nu k, -\nu k$ where ν is a square root of unity, i.e., $\nu^2 \equiv 1 \pmod{N}$. Now take $m \in \mathbb{Z}_N$ and assume $\gcd(m, N) = 1$. (Note that if we are able to compute $0 < m < N$ with $\gcd(m, N) \neq 1$, then we would have found a factor of N). Thus, compute $m^2 \pmod{N}$ and feed $\mathcal{A}$ with that input. If $\mathcal{A}$ outputs a value equal to m or $-m$, disregard and start again (this happens with probability $1/2$). Otherwise, if the output is $\hat{m}$ we have that $\gcd(m - \hat{m}, N) \neq 1$, so it must be a factor of N.

The Diffie–Hellman key agreement protocol was the first published practical method for establishing a shared secret-key over an authenticated (but not private) communication channel without using an a priori shared secret. The published description of this protocol is recognized as the paper that started *asymmetric* or *public-key* cryptography. Notwithstanding this, about twenty years later, it was publicly disclosed that asymmetric key algorithms had been developed a few years before by James Ellis, Clifford Cocks, and Malcolm Williamson at the Government Communications Headquarters in the UK (see a summary of their work on the subject in [Ell97]). In an asymmetric setting different keying material is (publicly) distributed to users depending on their role. As a result, key management can become simpler, resulting in cheaper and more flexible schemes. Moreover, the development of different dedicated cryptographic protocols beyond encryption (signature, authentication, identification, etc.) becomes possible. The advent of public-key cryptography led to a variety of new methods useful in many different scenarios, including the use of cryptographic techniques outside military applications. Today, symmetric-key and public-key cryptography complement each other and cryptographic constructions try to exploit the particular strengths of each.

3.2   Modern cryptology

The way information is created, stored, and transmitted today has little in common with information processing as it was known fifty years ago. Cryptography is used every day in modern society, by the average citizen, often obliviously. At the same time people are becoming more aware of the dangers of leaving sensitive data exposed, and ensuring the confidentiality of data has become a concern not only to corporations and governments, but also to individuals. The design of different cryptographic tools to meet a wide variety of newly demanded services is an ongoing process. An incomplete list of generic goals to pursue is the following:

Confidentiality. Secure message transmission and message storage is still one of the main goals of cryptographic constructions, though nowadays multiple scenarios are considered beyond the "plain" two-party communication we discussed so far. We may consider different sender sets that aim at interacting with dedicated receiver sets (e.g. *threshold schemes, broadcast schemes*), or message transmission which takes into account certain features of the input plaintext in order to enable an efficient management of encrypted data (*attribute-based encryption, searchable encryption*). We may use public-key encryption schemes in a setting in which public keys are not previously distributed, but linked to the users' identities (*identity-based encryption*), or schemes that not only hide the message, but also the keying material used to secure it (*key-private encryption*).

Further, the preservation of data confidentiality can also be a concern when carrying out cooperative computations with non-trusted partners; the research area dealing with this type of problem is *multiparty computation*.

Authentication. When exchanging information over an insecure channel one would like to authenticate both information itself and any entities involved in the communication. Examples of cryptographic constructions geared towards this goal are *signature schemes* and *identification schemes*.

Data integrity. Preventing unauthorized alteration of data is crucial whenever we are exchanging or storing information. It is often achieved using dedicated encryption schemes (cf. *non-malleable encryption schemes*, see Definition 4.6). Hash functions are a basic building block of integrity-preserving cryptographic design. These functions *condense* information, in the sense that they transform input bitstrings into (possibly much shorter) bitstrings of fixed length. As a result, they are "highly" non-injective, but designed in such a way that this lack of injectivity is hard to exploit computationally (we will learn more about this in Sections 4.3 and 8.1).

Non-repudiation. When data are processed, there is often the need to guarantee that the involved entities will not be able to deny their actions. Non-repudiation techniques allow for many applications in asynchronous networks (like the Internet)—such as e-auctions, e-bets, voting protocols, etc. A common cryptographic tool serving as a building block for schemes providing non-repudiation is known as a *commitment scheme*.

Let us take a look at basic examples for a signature scheme and a commitment scheme.

Example 3.2 [RSA signature]
In the same paper where RSA encryption was proposed [RSA78], a digital signature protocol was suggested with basically the same strategy as for encryption. More specifically, now Alice (the signer) computes N a product of two primes p, q, with $\log_2 p \approx \log_2 q$, and e a unit from $\mathbb{Z}_{\varphi(n)}$. Again, (N, e) is the public key, while e's multiplicative inverse, d, is the signing key that she will keep secret.

1. Alice signs a message $m \in \mathbb{Z}_N$ as $\sigma = m^d \pmod{N}$, and sends Bob the message-signature pair (m, σ).

2. Bob accepts the signature if $\sigma^e = m^{ed} = m$ in $\mathbb{Z}_N$, and otherwise rejects Alice's authorship.

While being intuitive, this basic scheme has a major security flaw. It is *existentially forgeable*, for anyone can obtain a valid signature pair (m, σ) from the public information by setting, for any $\sigma \in \mathbb{Z}_N$, $m = \sigma^e \pmod{N}$. Almost 20 years later, Bellare and Rogaway proposed a modification of this scheme known as RSA-FDH [BR96] that made use of a full domain hash function

$\mathcal{H}$; the idea is to pre-process m using the hash function and then signing this processed message $\mathcal{H}(m)$. As the function $\mathcal{H}$ is public, verification is as simple as in the standard RSA signature. As we will see in Chapter 10 the security of this scheme can be proved in the *random oracle model*, which we will see in Section 4.3. ▯

Example 3.3 [Commitment]
Take a cyclic group G of prime order q within $\mathbb{F}_p^*$, where p and q are odd primes. Select two elements $h, g \neq 1$ from G. As g is a generator of G, there must be a value l so that $h = g^l \pmod{p}$. This exponent l is supposed to be an unknown value, whereas p, q, h, and g are public information for all potential users. To commit to a value $v \in \mathbb{Z}_q$, Alice selects uniformly at random an exponent $r \in \mathbb{Z}_q$ and computes $C = g^v h^r \pmod{p}$. Anyone from the authorized group holding C will have a "receipt" for the value v without learning anything about this value (the scheme is *perfectly secret*, as C is consistent with every different v). However, Alice is bound to her initial choice, for (assuming she cannot retrieve the value l from g and h) she can only prove having constructed C by the above method disclosing the pair (v, r), and will not be able to compute an r-value consistent with any other $\hat{v} \in \mathbb{Z}_q$. ▯

Before the 1980s, cryptographic constructions were hardly evaluated in a rigorous way, and there was a (seemingly endless) cycle of design/cryptanalysis achievements. Proposals were improved heuristically using increasingly sophisticated tools (first, in a "mechanical" sense, later, in a mathematical sense), and yet, it was not uncommon that after some time security flaws were identified. Modern cryptography aims at designing cryptographic tools with clear security guarantees and limitations. The way in which this is achieved may be summarized in three principles (following [KL07]):

1. *Formulation of exact definitions.* It is crucial to define in a precise way the cryptographic goal the designed tool is geared towards, as well as the type of adversaries that will be considered and the security properties we aim at.

2. *Reliance on precise assumptions.* We may assume that the adversary has a certain (typically computational) limitation. This should be made explicit and such assumptions should be as minimal as possible.

3. *Provision of rigorous proofs of security.* Once the above principles are fulfilled, the claimed security properties should be rigorously proved under the assumptions we made explicit.

Working in accordance with these principles does not guarantee that a proposed scheme is secure. Assumptions—and even proofs—can be wrong, and there may be adversarial capabilities that we fail to model. Despite these imperfections, these principles offer valuable guidelines, in that they ensure that a (hopefully significant) class of attacks can be ruled out unless (hopefully plausible) assumptions fail.

Principle two of the above list asserts the importance of grounding cryptographic constructions on sound assumptions that are moreover explicitly formulated, but where do these assumptions come from? If we focus on the public-key setting, most widely deployed tools base their security on number theoretical problems (like factoring integers and computing discrete logarithms), but these are not the only problems under consideration, and it is desirable to avoid putting all eggs in the same basket. In Figure 3.3 we list some examples of problems from which cryptographic assumptions have been derived, and name some of the related constructions (a few of which will be discussed later in this book).

area	problem assumed to be hard	construction
number theory	factoring	RSA, Rabin
	discrete logarithm (dlog) in $\mathbb{F}_q$	Diffie–Hellman
	quadratic residuosity	Paillier
	elliptic curve analogue of dlog	IB-cryptography
combinatorics & algebra	knapsack	Merkle–Hellman
	non-linear systems of equations	Matsumoto–Imai
	ideal membership testing	Polly Cracker
group theory	word problem	Wagner–Magyarik
	conjugacy problem	braid schemes
	(log. signature) factoring	MST, PGM

FIGURE 3.3: Where do cryptographic assumptions come from?

This book is focusing on cryptographic constructions that use group theoretical problems as a base; namely, schemes which rely on the impossibility of completing in reasonable time computational tasks linked to group theory. Unfortunately, several proposals in this area have not taken into account the above principles, which, as we will see, have had an influence on their subsequent history.

3.3   Summary and further reading

In this chapter we have presented a brief (and very incomplete!) history of cryptology. We have seen that heuristic constructions were dominant in cryptographic design until the end of twentieth century. Main goals and design principles behind modern cryptology have been commented on, and we saw examples of cryptographic schemes for key agreement, encryption, signature, and commitment.

The reader interested in historical ciphers is likely to enjoy the books by Singh [Sin99] and Schneier [Sch04] as well as the paper of D. Kahn on cryptology during World Wars I and II [Kah80]. There are many good books for getting into modern cryptography from an undergraduate level, and we just mention here three of them [Sma03, Vau06, KL07].

3.4 Exercises

Exercise 17 *In the description of the RSA encryption scheme, we wrote that $\log_2 p$ should be approximately equal to $\log_2 q$, i. e., the secret prime factors should have comparable size. Is it desirable to choose the prime numbers p and q to be very close to each other?*

Exercise 18 *[Hill cipher] The Hill cipher was defined in the 1920s and is a classical cipher that can be described as follows: For fixed message length n, the encryption key is an $n \times n$ matrix E which is invertible modulo 26. Given a plaintext, remove all spaces and punctuation symbols, convert all letters into uppercase and substitute them with numbers in $\mathbb{Z}_{26}$ (set, e. g., $A = 0$, $B = 1$, etc.).*

Let us assume the plaintext length is a multiple of n, and parse this plaintext into k blocks of size n, say $v_1, \ldots, v_k$, which can be seen as column vectors in $\mathbb{Z}_{26}^n$. Multiply the two matrices E and $(v_1, \ldots, v_k)$ to obtain a ciphertext matrix $(c_1, \ldots, c_k)$ that may be transformed into a sequence of letters reverting the initial substitution. This is a secret-key scheme—the secret key can be easily derived by computing $E^{-1} \in \mathrm{GL}_n(\mathbb{Z}_{26})$.

Now assume you have intercepted a ciphertext sequence

$$EWLSZCKSPZKRUY$$

which you know corresponds to the message

$$SECRETSANDLIES.$$

Would you be able to retrieve the encryption key?

Exercise 19 *Prove that in the RSA encryption scheme deriving d and $\phi(N)$ from N is comparable in difficulty to finding the prime factorization of N.*

Exercise 20 *Complete the proof of Theorem 3.1.*

Exercise 21 *[Merkle–Hellman Cryptosystem] Let Alice (the receiver) select a super-increasing sequence of positive integers $w = (w_1, \ldots, w_n)$. This means that*

$$w_{i+1} > \sum_{j=1}^{i} w_j \text{ for } i = 1, \ldots, n-1.$$

Further, she chooses q, r two relatively prime integers with $q > \sum_{i=1}^{n} w_i$ and derives a sequence $\beta = (\beta_1, \ldots, \beta_n)$ with $\beta_i = rw_i \pmod{q}$, for $i = 1, \ldots, n$.

Alice publishes β, while keeping w, q, and r secret. If Bob wants to send her an n-bit message $b_0, \ldots, b_{n-1}$, he encrypts this message as

$$c = \sum_{i=1}^{n} b_i \beta_i.$$

Describe Alice's decryption strategy, and try to identify sound assumptions on which the security of this scheme could rely.

Exercise 22 *[Schnorr signature] Let G be a cyclic group of prime order q with publicly known generator g, and let $\mathcal{H} : \{0,1\}^* \longrightarrow \mathbb{Z}_q$ be a hash function. We may define a signature scheme as follows:*

Alice (the signer) chooses uniformly at random a value $x \in \mathbb{Z}_q^*$ and publishes g^{-x}. For signing a message $m \in \{0,1\}^*$, Alice selects a random exponent $k \in \mathbb{Z}_q^*$, encodes the group element $r = g^k$ is as a bitstring, and computes the values $h = \mathcal{H}(m, r)$ and $s = k + xh \pmod{q}$. The triplet (r, h, s) serves as signature for m and is, along with m, sent to the verifier. Define a verification procedure and list two assumptions that seem reasonable to demand in order to establish the security of this scheme.

Exercise 23 *Prove that the problem of breaking the binding property of the commitment scheme from Example 3.3 reduces indeed to computing discrete logarithms to base g.*

Part II

Public-Key Encryption

Chapter 4

Provable security guarantees

So far, our discussion of public-key encryption has been very informal, and we made no serious attempt to define what we mean by a "secure" public-key encryption scheme. In this chapter we establish some terminology that enables us to formalize security requirements and to establish strong provable guarantees for public-key encryption schemes under clearly defined assumptions. Furthermore, we discuss a first generic construction to derive secure asymmetric encryption from basic tools (trapdoor one-way functions and random oracles) and comment on an idea for developing secure constructions using certain factorization sequences of finite groups (logarithmic signatures).

4.1 Public-key encryption revisited

The intuitive approach we used so far to describe and discuss encryption schemes is helpful for conveying basic ideas. There is a significant danger, however, of overlooking relevant attacks that could have been avoided by a more careful analysis.

Example 4.1

The basic RSA scheme in Section 3.1.2 is deterministic, i. e., if we encrypt the same plaintext twice with the same public key, then we obtain exactly the same ciphertext. Consequently, if the set of possible plaintexts is small, then an adversary may compute a table containing all possible plaintexts along with their (unique!) encryption under a given public key. With such a table, decrypting an eavesdropped message reduces to a simple table look-up. ▯

In regard to small message spaces, the ElGamal scheme from Section 3.1.2 appears to be more robust than RSA: The encryption algorithm[1] is probabilistic, and even for a fixed public key $pk = g^a$ and a fixed plaintext m, there

---

[1] As for a function we expect to always obtain the same value when evaluating it at the same argument, we do not use the term *encryption function*. Instead, we describe the process of encrypting a plaintext as applying a probabilistic algorithm to it.

are different possible ciphertexts $c = (g^r, pk^r \cdot m)$. Differing from the basic RSA scheme, for ElGamal the encryption algorithm involves the choice of a random element, and later on in this section we will see that for a suitable choice of parameters ElGamal offers quite strong provable security guarantees. However, it is not difficult to see that in the described form ElGamal fails to protect the integrity of a message.

Example 4.2
Suppose the adversary eavesdrops some ciphertext

$$c = (g^r, pk^r \cdot m) \in G \times G$$

that has been computed with the ElGamal encryption scheme to protect a plaintext message m. Then a simple multiplication of the right component of c with a group element $h \in G$ allows the adversary to compute

$$c' = (g^r, pk^r \cdot mh),$$

which decrypts to $m' = m \cdot h$. In other words, the adversary can produce a ciphertext m' that is "meaningfully related" to m. Depending on the application one has in mind—say m encodes an amount of money in a financial transaction that can be increased through multiplication with h—this may pose a substantial problem.   ⬚

To enable a more thorough treatment of public-key encryption, let us first clarify what we mean by this term. A public-key encryption scheme will be specified by a collection of algorithms that are parameterized by a *security parameter* $\ell \in \mathbb{N}$. Intuitively, ℓ reflects a key length, and to express that an event becomes "rather unlikely provided that the key size is chosen to be large enough," the concept of a negligible function is handy:

DEFINITION 4.1 [Negligible function]   *A function* $f : \mathbb{N} \longrightarrow \mathbb{R}_{\geq 0}$ *is said to be* negligible *if for every* $c \in \mathbb{R}$ *there exists an integer* k_0 *such that*

$$\forall k > k_0 : f(k) < \frac{1}{k^c} .$$

Typically we will use negligible functions to express that the probability for a certain type of attack to succeed is quickly approaching zero, when the key size is increased.

Example 4.3
Suppose we choose two ℓ-bit strings $s_0, s_1 \in \{0,1\}^\ell$ uniformly and independently at random. Then the probability of s_0 being equal to s_1 is

$$\Pr(s_0 = s_1) = \frac{1}{2^\ell} ,$$

which (taken for a function in ℓ) is negligible. □

Having said this, let us define what we mean by a public-key encryption scheme.

DEFINITION 4.2 [Public-key encryption scheme]   A public-key encryption scheme *is given by a triple of polynomial time algorithms* $(\mathcal{K}, \mathcal{E}, \mathcal{D})$ *along with a family* $\{M_\ell\}_{\ell \in \mathbb{N}}$ *of message spaces. More specifically, we have the following:*

$\mathcal{K}$: *a probabilistic* key generation *algorithm that on input a security param-eter* $1^\ell = \underbrace{1 \ldots 1}_{\ell \text{ times}}$ *outputs a pair* (pk, sk). *We refer to pk as a* public key *and to sk as a* secret key.

$\mathcal{E}$: *a probabilistic algorithm that on input a public key pk and a plaintext* $m \in M_\ell$ *outputs a ciphertext c.*

$\mathcal{D}$: *a deterministic algorithm that on input a secret key sk and a candidate ciphertext c either outputs a plaintext* $m \in M_\ell$ *or a dedicated error symbol* $\perp \notin \bigcup_{\ell \in \mathbb{N}} M_\ell$.

For $\mathcal{E}$ and $\mathcal{D}$ we will write the first argument as an index, i. e., we set

$$\mathcal{E}_{pk}(m) = \mathcal{E}(pk, m) \text{ and}$$
$$\mathcal{D}_{sk}(c) = \mathcal{D}(sk, c).$$

Moreover, for key pairs (pk, sk) generated by the key generation algorithm $\mathcal{K}$, we require that the event

$$\exists m \in M_\ell : \mathcal{D}_{sk}(\mathcal{E}_{pk}(m)) \neq m$$

occurs with negligible probability only.[2]

The last condition reflects the idea that, with the possible exception of a few "bad keys," decrypting a ciphertext with the correct secret key must always yield the original plaintext. Allowing such a tiny set of bad keys in the definition gives us the option to use an efficient key generation algorithm which in theory could sometimes fail, but the probability of this actually happening is too small to be a real concern.

It is important to note that Definition 4.2 makes no statement about secu-rity yet. To make the intuitive idea of a secure public-key encryption scheme more precise, first we have to decide about the attacks we want to consider— and we will have to formalize them properly. A meaningful security definition

---

[2]Note again that, as is often the case, probability is considered as a function of the security parameter ℓ.

should cover a wide range of adversarial strategies and at the same time be simple enough to work with it in proofs. Identifying good security definitions is a non-trivial task; it actually is an active area of cryptographic research. For instance, when implementation-dependent aspects like power consumption, running time and electromagnetic emanation are to be taken into account, finding adequate theoretical security models is not an easy task. Common security definitions typically restrict to a large class of "structural" attacks and do not consider adversarial strategies that rely on characteristics of a specific implementation platform. This does not imply that attacks outside the chosen formal model are irrelevant, so it is important to be aware of the limitations of the guarantees that are established by a proof.

From a practical point of view, it is desirable that we can characterize the security level of an encryption scheme for specific key lengths—for instance, we might want to know how much of an effort it takes for an attacker to find the plaintext underlying an ElGamal encryption based on the multiplicative group $\mathbb{F}_{2^{199}}^*$ in a particular representation. To simplify the analysis, it can be convenient, however, to define formal security notions asymptotically and to postpone the (important) discussion of specific values of the security parameter to a later stage.

Different approaches can be taken to express mathematically what we mean by a "secure" public-key encryption scheme, and it is interesting to note that seemingly different paradigms turned out to characterize the same class of public-key encryption schemes. In the sequel we follow one of the most common approaches and characterize each security notion in terms of a game played by the adversary. We then try to show that under certain (hopefully plausible) assumptions for any efficient algorithm the probability of winning the game in question is negligible.

4.2 Characterizing secure public-key encryption

In order to be able to *prove* statements about the security of a public-key encryption scheme, we should specify what we mean by a successful attack. To this aim, we detail

- the security goal we want to achieve and

- the tools that are available to the adversary.

A first obvious security goal is that the adversary should not be capable of recovering a plaintext message from the ciphertext and a public key alone— provided that the plaintexts are distributed appropriately.

DEFINITION 4.3 [One-wayness]   *A public-key encryption scheme* $(\mathcal{K}, \mathcal{E}, \mathcal{D})$ *is* secure against chosen plaintext attacks in the sense of one-wayness (OW-CPA) *if for every probabilistic polynomial time algorithm $\mathcal{A}$ the following holds: On input the security parameter 1^ℓ, a public key pk (with (pk, sk) output by $\mathcal{K}(1^\ell)$), and a ciphertext c obtained as $\mathcal{E}_{pk}(m)$ with a uniformly at random chosen plaintext $m (\in M_\ell)$, the probability of $\mathcal{A}$ outputting m differs at most negligibly from $1/|M_\ell|$, where $|M_\ell|$ is the size of the message space M_ℓ.*

REMARK 4.1   In the above definition, one could alternatively impose that the probability of $\mathcal{A}$ outputting the correct m is negligible. As discussed in Exercise 26, such a definition has some undesirable consequences, however.
⬜

In Definition 4.3, choosing the message $m \in M_\ell$ uniformly at random may seem artifical, and one could consider different probability distributions on the plaintext space. However, one-wayness in the above (rather weak) sense turns out to be a good starting point to establish much stronger security guarantees. Before defining such stronger security requirements, we introduce some notation which enables a more compact formulation of definitions like the above: For an algorithm $\mathcal{A}$ we write

$$y \leftarrow \mathcal{A}(x)$$

to indicate that $\mathcal{A}$ on input x outputs y. In particular, y depends on the random choices made by $\mathcal{A}$, and from

$$y_0 \leftarrow \mathcal{A}(x); \ y_1 \leftarrow \mathcal{A}(x)$$

we can in general *not* conclude that $y_0 = y_1$. For a finite set S, we use the notation

$$s \leftarrow S$$

to indicate that s is chosen from S uniformly at random. With this notation, our definition of one-wayness can be phrased in terms of the experiment in Figure 4.1: For a given public-key encryption scheme $(\mathcal{K}, \mathcal{E}, \mathcal{D})$ and a given probabilistic polynomial time algorithm $\mathcal{A}$ we define the function

$$\mathbf{Succ}_{\mathcal{A}^{\text{ow-cpa}}, (\mathcal{K}, \mathcal{E}, \mathcal{D})}(\ell) = \Pr[\mathbf{Exp}_{\mathcal{A}, (\mathcal{K}, \mathcal{E}, \mathcal{D})}(1^\ell) = 1],$$

where the probability is over all random choices involved (i.e., sampling from M_ℓ and the random choices of $\mathcal{A}$, $\mathcal{K}$, and $\mathcal{E}$). We say that $(\mathcal{K}, \mathcal{E}, \mathcal{D})$ is *secure in the sense of OW-CPA* if for every probabilistic polynomial time adversary $\mathcal{A}$ the advantage

$$\mathbf{Adv}^{\text{ow-cpa}}_{\mathcal{A}, (\mathcal{K}, \mathcal{E}, \mathcal{D})}(\ell) = \left| \mathbf{Succ}_{\mathcal{A}^{\text{ow-cpa}}, (\mathcal{K}, \mathcal{E}, \mathcal{D})}(\ell) - \frac{1}{|M_\ell|} \right|$$

$$
\begin{array}{|l|}
\hline
\mathbf{Exp}_{\mathcal{A},(\mathcal{K},\mathcal{E},\mathcal{D})}\text{OW-CPA}(1^{\ell}) \\
(pk, sk) \leftarrow \mathcal{K}(1^{\ell}) \\
m \leftarrow M_{\ell} \\
c \leftarrow \mathcal{E}_{pk}(m) \\
d \leftarrow \mathcal{A}(1^{\ell}, pk, c) \\
\text{if } d = m \\
\quad \text{then return } 1 \\
\quad \text{else return } 0 \\
\hline
\end{array}
$$

FIGURE 4.1: One-wayness under chosen plaintext attacks.

is negligible.

The security guarantees offered by OW-CPA are rather modest: For a scenario where plaintexts do not follow the (uniform) distribution prescribed in the definition, no guarantees are made. Further on, it is only imposed that recovering a *complete* plaintext is hard. The definition does not exclude a situation where parts of the plaintext can easily be recovered from a ciphertext and the public key. Here the notion of *ciphertext indistinguishability* we discuss next goes a significant step further.

As described in the experiment in Figure 4.2 we now allow the adversary $\mathcal{A}$ to choose two arbitrary plaintexts m_0 and m_1 from the appropriate message space M_{ℓ}, and he only has to decide whether a received ciphertext is an encryption of m_0 or an encryption of m_1. To avoid trivialities with public-key encryption schemes where plaintexts can vary in length we impose that m_0 and m_1 have the same length. Differing from the experiment for OW-CPA security, now we have two separate phases, and in Figure 4.2 this materializes as an auxiliary value —find or guess— being part of the adversary's input. The output s in the first (find-)phase in the experiment is state information, where $\mathcal{A}$ can store values which might be of use for the second (guess-)phase.

DEFINITION 4.4 [Ciphertext indistinguishability: IND-CPA]
For a public-key encryption scheme $(\mathcal{K}, \mathcal{E}, \mathcal{D})$ and an algorithm $\mathcal{A}$ consider the experiment in Figure 4.2. We refer to $(\mathcal{K}, \mathcal{E}, \mathcal{D})$ as secure against chosen plaintext attacks in the sense of indistinguishable encryptions (IND-CPA) *if for every probabilistic polynomial time algorithm $\mathcal{A}$ the advantage*

$$
\mathbf{Adv}^{\text{ind-cpa}}_{\mathcal{A},(\mathcal{K},\mathcal{E},\mathcal{D})}(\ell) = |2 \cdot \Pr[\mathbf{Exp}_{\mathcal{A},(\mathcal{K},\mathcal{E},\mathcal{D})}\text{IND-CPA}(1^{\ell}) = 1] - 1|
$$

is negligible.

For the ElGamal encryption scheme, which we have seen in Section 3.1.2 already, it is not hard to see that it can provide security in the sense of IND-CPA. Looking at the proof of this guarantee, we see that for establishing this result, we actually rely on the so-called decisional Diffie–Hellman (DDH)

$$
\begin{array}{|l|}
\hline
\textbf{Exp}_{\mathcal{A},(\mathcal{K},\mathcal{E},\mathcal{D})}\text{IND-CPA}(1^\ell) \\
(pk, sk) \leftarrow \mathcal{K}(1^\ell) \\
(m_0, m_1, s) \leftarrow \mathcal{A}(1^\ell, \mathsf{find}, pk) \\
b \leftarrow \{0,1\} \\
c \leftarrow \mathcal{E}_{pk}(m_b) \\
d \leftarrow \mathcal{A}(1^\ell, \mathsf{guess}, s, c) \\
\text{if } d = b \text{ and } |m_0| = |m_1| \\
\quad \text{then return } 1 \\
\quad \text{else return } 0 \\
\hline
\end{array}
$$

FIGURE 4.2: Indistinguishable encryptions under chosen plaintext attacks.

assumption. Informally, the DDH assumption for a cyclic group generated by g says only that tuples of the form $(g, g^\alpha, g^\beta, g^{\alpha \cdot \beta})$ with random exponents α, β cannot be efficiently distinguished from tuples $(g, g^\alpha, g^\beta, g^\gamma)$ with α, β, γ being chosen at random. Thus, proving IND-CPA security under a DDH assumption does not say that a successful adversary can compute discrete logarithms in the underlying group.

Assumption 4.1 (Decisional Diffie–Hellman assumption)
Let $\{G_\ell\}_\ell$ be a family of, multiplicatively written, finite cyclic groups along with a polynomial time algorithm $\mathcal{G}$ that on input 1^ℓ outputs a generator g_ℓ for G_ℓ (along with polynomial time algorithms for computing the product and for deciding equality of elements in G_ℓ). The decisional Diffie–Hellman (DDH) assumption states that for every probabilistic polynomial time algorithm $\mathcal{A}$ the advantage

$$
\mathbf{Adv}_{\mathcal{A},\mathcal{G}}^{\mathrm{ddh}}(\ell) = |2 \cdot \Pr(\mathbf{Exp}_{\mathcal{A},\mathcal{G}}\text{DDH}(1^\ell) = 1) - 1|
$$

is negligible, with the experiment $\mathbf{Exp}_{\mathcal{A},\mathcal{G}}\text{DDH}$ being specified in Figure 4.3.

In abuse of terminology, we often speak of the "DDH assumption for the group G," assuming that the family of groups G_ℓ to which the group $G = G_{\ell_0}$ belongs and the choice of the generator are clear. Typical examples include prime order subgroups of finite fields $\mathbb{F}_q^*$ and prime order subgroups of elliptic curves $E(\mathbb{F}_q)$.

PROPOSITION 4.1
Suppose that the DDH assumption holds in $\{G_\ell\}_\ell$. Then the ElGamal encryption scheme with message space $\mathcal{M}_\ell = G_\ell$ as specified in Figure 4.4 is secure in the sense of IND-CPA.

PROOF Suppose $\mathcal{A}$ is an algorithm that violates IND-CPA security. Then we can derive an algorithm $\mathcal{B}$ to solve the DDH problem in the underlying

$$
\begin{array}{l}
\mathbf{Exp}_{\mathcal{A},\mathcal{G}}\text{DDH}(1^\ell) \\
\quad g \leftarrow \mathcal{G}(1^\ell) \\
\quad \alpha \leftarrow \{1, \ldots, \text{ord}(g)\} \\
\quad \beta \leftarrow \{1, \ldots, \text{ord}(g)\} \\
\quad b \leftarrow \{0, 1\} \\
\quad \text{if } b = 0 \\
\qquad \text{then } \gamma \leftarrow \{1, \ldots, \text{ord}(g)\} \\
\qquad \text{else } \gamma = \alpha \cdot \beta \\
\quad d \leftarrow \mathcal{A}(1^\ell, g, g^\alpha, g^\beta, g^\gamma) \\
\quad \text{if } d = b \\
\qquad \text{then return } 1 \\
\qquad \text{else return } 0
\end{array}
$$

FIGURE 4.3: Decisional Diffie–Hellman problem.

Key generation $\mathcal{K}$: on input the security parameter ℓ, choose a generator $g \leftarrow G_\ell$ and an exponent $\alpha \leftarrow \{1, \ldots, \text{ord}(g)\}$. The secret key is $sk = \alpha$, and the public key is $pk = (g, g^{sk})$.

Encryption $\mathcal{E}$: to encrypt a message $m \in G_\ell$ under public key (g, g^α), choose an exponent $\beta \leftarrow \{1, \ldots, \text{ord}(g)\}$ and output $c = (g^\beta, (g^\alpha)^\beta \cdot m)$.

Decryption $\mathcal{D}$: to decrypt a ciphertext $c = (g^\beta, h)$ under secret key $sk = \alpha$, compute $\left((g^\beta)^\alpha\right)^{-1} \cdot h$.

FIGURE 4.4: ElGamal encryption scheme.

group G_ℓ: $\mathcal{B}$ obtains as input a tuple $(g, g^\alpha, g^\beta, g^\gamma)$ and runs a simulation of $\mathcal{A}(1^\ell, \mathsf{find}, \cdot)$ with input the public key (g, g^α). On receiving messages m_0, m_1 from $\mathcal{A}$, $\mathcal{B}$ chooses a random bit $b \leftarrow \{0, 1\}$ and sends the challenge ciphertext $(g^\beta, g^\gamma \cdot m_b)$, along with the state information s received at the end of the find stage, to a simulation of $\mathcal{A}(1^\ell, \mathsf{guess}, \cdot)$. Whenever the latter identifies b correctly, then $\mathcal{B}$ returns 1, i.e., claims that the challenge $(g, g^\alpha, g^\beta, g^\gamma)$ satisfies $\gamma = \alpha \cdot \beta$. Otherwise $\mathcal{B}$ returns 0, i.e., suspects to have received a random exponent γ.

To compute the advantage of $\mathcal{B}$, we look at both possible cases for the DDH challenge:

- If the random value b in the DDH experiment is equal to 1, the challenge received by $\mathcal{B}$ satisfies $\gamma = \alpha \cdot \beta$ and $\mathcal{B}$ correctly identifies b if and only if $\mathcal{A}$ solves the IND-CPA challenge correctly.

- If the random value b in the DDH experiment is equal to 0, $\mathcal{B}$ sends a ciphertext $(g^\beta, g^\gamma \cdot m_b)$ to $\mathcal{A}(1^\ell, \mathsf{guess}, \cdot)$ where γ is chosen uniformly at random. In this case, $\mathcal{A}(1^\ell, \mathsf{guess}, \cdot)$ guesses correctly with probability $1/2$, as $g^\gamma \cdot m_b$ is just a uniformly at random chosen group element.

Denoting by $\mathbf{Succ}_{\mathcal{A}}^{\text{ind-cpa}}$ the success probability in solving the ind-cpa challenge, and, in the same way, by $\mathbf{Succ}_{\mathcal{B}}^{\text{ddh}}$ the success probability in solving the ddh challenge, we obtain

$$\mathbf{Succ}_{\mathcal{B}}^{\text{ddh}} = \frac{1}{2} \cdot \mathbf{Succ}_{\mathcal{A}}^{\text{ind-cpa}} + \frac{1}{2} \cdot \frac{1}{2},$$

and therefore $\mathbf{Adv}_{\mathcal{B},G}^{\text{ddh}} = \left| \mathbf{Succ}_{\mathcal{A}}^{\text{ind-cpa}} + \frac{1}{2} - 1 \right| = \frac{1}{2} \cdot \mathbf{Adv}_{\mathcal{A},\text{ElGamal}}^{\text{ind-cpa}}.$ As the advantage for a probabilistic polynomial time $\mathcal{B}$ is by assumption negligible, we recognize $\mathcal{A}$'s advantage in attacking the ciphertext indistinguishability of ElGamal as negligible, too. □

A security reduction as in the proof of Proposition 4.1 is sometimes referred to as *tight* to indicate that the advantage in violating the desired security guarantee (in this case IND-CPA security) and the advantage in violating the underlying hardness assumption differ by no more than a "small" constant factor.

In a scenario where an adversary $\mathcal{A}$ has access to the public key and is only passively eavesdropping communication, security in the sense of IND-CPA is a rather strong guarantee. However, suppose that $\mathcal{A}$ has, say prior to receiving a ciphertext c, the possibility to obtain decryptions of some ciphertexts of its choosing. Such an attack is sometimes called a "lunch time attack"—with the idea that $\mathcal{A}$ uses the lunch break (corresponding to the find-phase) of a victim to access a (tamper proof) decryption device. To model such *chosen ciphertext attacks*, we equip the adversary in the experiment from Figure 4.2 with a decryption oracle that knows the secret key and runs the decryption algorithm on any input of the adversary's choosing. Note that in particular the adversary is allowed to submit invalid inputs to the decryption oracle, which is often more useful than one may naively suspect. Furthermore, we may consider a stronger notion in which the access to this oracle is not limited to the find-phase. To avoid a situation where for trivial reasons no secure public-key encryption scheme can exist, we have to make one obvious restriction: the adversary is not allowed to query the challenge ciphertext to the decryption oracle.

DEFINITION 4.5 [Ciphertext indistinguishability: IND-CCA]

For a public-key encryption scheme $(\mathcal{K}, \mathcal{E}, \mathcal{D})$ and an algorithm $\mathcal{A}$ consider the experiment in Figure 4.2. We refer to $(\mathcal{K}, \mathcal{E}, \mathcal{D})$ as secure against non-adaptive chosen ciphertext attacks in the sense of indistinguishable encryptions (IND-CCA1) if for every probabilistic polynomial time algorithm $\mathcal{A}$ the advantage

$$\mathbf{Adv}_{\mathcal{A},(\mathcal{K},\mathcal{E},\mathcal{D})}(\ell) = |2 \cdot \Pr[\mathbf{Exp}_{\mathcal{A},(\mathcal{K},\mathcal{E},\mathcal{D})}\text{IND-CCA1}(1^\ell) = 1] - 1| \qquad (4.1)$$

is negligible, even if $\mathcal{A}$ has during the find-*phase of the experiment, i. e., before learning the challenge ciphertext c, unrestricted access to a decryption oracle $\mathcal{O}_{\mathcal{D}}$ running the decryption algorithm $\mathcal{D}_{sk}(\cdot)$.*

If the advantage (4.1) remains negligible, if $\mathcal{A}$ has access to $\mathcal{O}_{\mathcal{D}}$ during both the find- *and the* guess-*phase—subject only to the restriction that $\mathcal{A}$ must not query the challenge ciphertext c in the* guess-*phase—then we refer to $(\mathcal{K}, \mathcal{E}, \mathcal{D})$ as* secure against adaptive chosen ciphertext attacks in the sense of indistinguishable encryptions (IND-CCA2).

The security guarantee captured by Definition 4.5 is quite strong, and when aiming at the confidentiality of plaintexts, such a guarantee seems sufficient for many applications. As we will see later, under suitable assumptions public-key encryption schemes meeting this level of security can be realized efficiently. By now, security in the sense of IND-CCA2 is considered a standard requirement for public-key encryption schemes.

REMARK 4.2   It is common practice to omit the number 2 at the end of IND-CCA2 and just write IND-CCA when referring to ciphertext indistinguishability under adaptive chosen ciphertext attacks. We will follow this convention throughout this book.   □

Despite offering seemingly strong guarantees, it is worthwhile to ask, if there are requirements other than ciphertext indistinguishability that one may like to impose on a public-key encryption scheme.

Example 4.4
Consider two companies that are competing for a particular contract. Company A encrypts its offer m under the public key of the prospective customer. Suppose the competing company B eavesdrops the ciphertext c encrypting A's offer. Now it is desirable that B cannot modify the ciphertext c in such a way that B obtains a ciphertext c' encoding a more favorable offer m', e.g., a price that is 10% below A's offer.   □

Note that in this example the "adversary" B does not aim at learning the plaintext (A's offer), but rather at producing a *ciphertext* that encrypts a (possibly unknown) plaintext m' that is in a known relation to m. To formalize such a security requirement, we proceed similarly as in the definition of ciphertext indistinguishability and set up a game with a find- and a guess-stage. To formalize the idea of an adversary coming up with a "meaningfully related ciphertext," we set up the game a successful adversary has to win along the following lines:

find-**phase.** On input of a public key, the adversary outputs a description of a probabilistic polynomial time sampling algorithm $\mathcal{M}$ which samples plaintexts from (all or parts of) the message space M_ℓ, plus any additional information extracted from the find stage he wishes to communitate to the **guess** stage of the experiment (value s from Figure 4.5). The idea is that this sampling algorithm outputs messages whose encryptions can be modified by $\mathcal{A}$ in a meaningful manner.

challenge phase. Now the sampling algorithm $\mathcal{M}$ is executed twice to produce plaintexts m_0, m_1, and the encryption algorithm is applied to m_1, resulting in a challenge ciphertext c_1 which is handed over to the adversary $\mathcal{A}$. The second plaintext message m_0 remains hidden from the adversary and will later serve as a tool to measure the success of $\mathcal{A}$.

guess-phase. On input the challenge ciphertext c_1 and state information (output at the end of the find-phase), $\mathcal{A}$ outputs a description of a deterministic polynomial time algorithm $\mathcal{R}$ along with a finite sequence of ciphertexts $c = [c_2, \ldots, c_r]$ of encryptions of plaintexts from the message space. We do not impose that $\mathcal{A}$ knows the underlying plaintexts m_i of the ciphertexts c_i, but for the adversary $\mathcal{A}$ to win, $\mathcal{D}_{sk}(c_i) \neq \perp$ must hold for all $i = 2, \ldots, r$. Further on, we impose that $c_1 \neq c_i$ for all $i = 2, \ldots, r$—this prevents that a simple replay of the challenge ciphertext is considered a valid attack.

On input r plaintexts from the message space, the algorithm $\mathcal{R}$ outputs either 0 or 1, i. e., $\mathcal{R}$ can be regarded as testing if a certain (r-ary) relation holds (output value 1) or not (output value 0). The adversary is considered successful if the probability of $\mathcal{R}(m_1, m_2, \ldots, m_r)$—for which we write in short $\mathcal{R}(m_1, m)$—to hold is non-negligibly different from the probability of $\mathcal{R}(m_0, m)(= \mathcal{R}(m_0, m_2, \ldots, m_r))$ to hold. This captures the idea that $\mathcal{A}$ can indeed relate the plaintext underlying the challenge ciphertext to other encrypted messages in a meaningful way.

The following definition uses the notation we introduced at the beginning of this section to describe this game more concisely.

DEFINITION 4.6 [Non-malleability: NM-CPA]
For a public-key encryption scheme $(\mathcal{K}, \mathcal{E}, \mathcal{D})$ and an adversary $\mathcal{A}$ consider the experiment in Figure 4.5. We refer to $(\mathcal{K}, \mathcal{E}, \mathcal{D})$ as secure against chosen plaintext attacks in the sense of non-malleability (NM-CPA) *if for every probabilistic polynomial time algorithm $\mathcal{A}$ and every polynomial $p(\ell)$ we have the following: if the running time of $\mathcal{A}$ and of each of the algorithms $\mathcal{M}$ and $\mathcal{R}$ output by $\mathcal{A}$ is bounded by $p(\ell)$, the advantage*

$$\mathbf{Adv}_{\mathcal{A},(\mathcal{K},\mathcal{E},\mathcal{D})}(\ell) = \Pr[\mathbf{Exp}_{0,\mathcal{A},(\mathcal{K},\mathcal{E},\mathcal{D})}\text{NM-CPA}(1^\ell) = 1] \qquad (4.2)$$

is negligible.

$$
\begin{aligned}
&\mathbf{Exp}_{\mathcal{A},(\mathcal{K},\mathcal{E},\mathcal{D})}\text{NM-CPA}(1^{\ell}) \\
&\quad (pk, sk) \leftarrow \mathcal{K}(1^{\ell}) \\
&\quad (\mathcal{M}, s) \leftarrow \mathcal{A}(1^{\ell}, \mathsf{find}, pk) \\
&\quad m_0 \leftarrow \mathcal{M}(1^{\ell}); \; m_1 \leftarrow \mathcal{M}(1^{\ell}) \\
&\quad c_1 \leftarrow \mathcal{E}_{pk}(m_1) \\
&\quad (\mathcal{R}, c) \leftarrow \mathcal{A}(1^{\ell}, \mathsf{guess}, s, c_1) \\
&\quad \text{if } \mathcal{R}(1^{\ell}, m_1, m) \neq \mathcal{R}(1^{\ell}, m_0, m) \\
&\qquad \text{then return } 1 \\
&\qquad \text{else return } 0
\end{aligned}
$$

FIGURE 4.5:   Non-malleability under chosen plaintext attacks.

It is not difficult to verify that the ElGamal encryption scheme offers little protection against this type of attack (see Exercise 28). On the other hand we know from Proposition 4.1 that ElGamal is secure in the sense of IND-CPA. So non-malleability and indistinguishability of ciphertexts turn out to be different requirements with the former not being implied by the latter. Interestingly, the situation changes when looking at stronger adversaries.

DEFINITION 4.7 [Non-malleability: NM-CCA]

For a public-key encryption scheme $(\mathcal{K}, \mathcal{E}, \mathcal{D})$ and an adversary $\mathcal{A}$ consider the experiment in Figure 4.5. We refer to $(\mathcal{K}, \mathcal{E}, \mathcal{D})$ as secure against chosen ciphertext attacks in the sense of non-malleability (NM-CCA1) if for every probabilistic polynomial time algorithm $\mathcal{A}$ and every polynomial $p(\ell)$ we have the following: if the running time of $\mathcal{A}$ and of each of the algorithms $\mathcal{M}$ and $\mathcal{R}$ output by $\mathcal{A}$ is bounded by $p(\ell)$, the advantage

$$
\mathbf{Adv}_{\mathcal{A},(\mathcal{K},\mathcal{E},\mathcal{D})}(\ell) = \Pr[\mathbf{Exp}_{\mathcal{A},(\mathcal{K},\mathcal{E},\mathcal{D})}\text{NM-CPA}(1^{\ell}) = 1] \tag{4.3}
$$

is negligible, even if $\mathcal{A}$ has during the find-phase of the experiment, i. e., before learning the challenge ciphertext c, unrestricted access to a decryption oracle $\mathcal{O}_{\mathcal{D}}$ running the decryption algorithm $\mathcal{D}_{sk}(\cdot)$.

If the advantage (4.2) remains negligible, if $\mathcal{A}$ has access to $\mathcal{O}_{\mathcal{D}}$ during both the find- and the guess-phase—subject only to the restriction that $\mathcal{A}$ must not query the challenge ciphertext c in the guess-phase—then we refer to $(\mathcal{K}, \mathcal{E}, \mathcal{D})$ as secure against adaptive chosen ciphertext attacks in the sense of non-malleability (NM-CCA2).

REMARK 4.3   As in the case of IND-CCA security, it is common to drop the number 2 at the end of NM-CCA2 and to speak simply of NM-CCA security. We will follow this convention throughout the book.   ▯

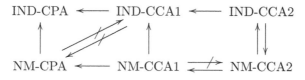

FIGURE 4.6: Relations among different security notions.

REMARK 4.4 Different security notions are coined for modeling adversaries with somewhat weaker oracles than the CCA one. A relevant example is the one capturing so-called *reaction attacks*. In that scenario, we assume the adversary has access to a *validity checking oracle* which on input of a public key and a (presumably valid) ciphertext will determine if the ciphertext is consistent with the input key and the encryption algorithm. Such an oracle, as we will have the chance to see later, is quite effective against a number of schemes (see [HGS99]).

□

Figure 4.6 illustrates implications among different security notions and in particular says that for adaptive chosen ciphertext attacks, imposing security against "malleability attacks" is equivalent to imposing security in the sense of ciphertext indistinguishability.[3] While IND-CCA is not the strongest known security notion for a public-key encryption scheme, it is fair to say that it is the most common one. Its equivalence to NM-CCA may be taken for an indication that a "natural" class of public-key encryption schemes is captured by this notion. This intuition is backed by the observation that a seemingly very different characterization of secure public-key encryption has turned out to be equivalent to IND-CCA security as well.

Before looking at how to build public-key encryption schemes meeting this level of security, let us present an example of a proposed public-key encryption scheme which has similarity with the ElGamal encryption scheme, but builds on a non-Abelian setting. The cryptanalytic successes against this proposal evidence that relying on an informal security analysis comes with a certain risk.

Example 4.5 [MOR encryption scheme]
The MOR scheme described here was published at the beginning of this century, and was proposed without a formal security analysis. One might be tempted to hope that the underlying non-Abelian structure allows for some-

---

[3]In Figure 4.6 arrows represent implications, so a directed path from security notion A to security notion B means that A implies B.

thing MORe than what is possible in the Abelian case, but, as cryptanalytic analysis evidenced, it unfortunately turned out not to be the case.

Let G be a finite non-Abelian group generated by $g_1, \ldots, g_n \in G$ and such that elements from G can be sampled uniformly at random. Let

$$\phi : G \longrightarrow \text{Aut}(G)$$
$$h \longmapsto \phi_h = (g \mapsto hgh^{-1})$$

be the homomorphism which maps each element in G to a corresponding inner automorphism. With this notation, Figure 4.7 describes key generation, encryption, and decryption in MOR.

Key generation $\mathcal{K}$: chooses $h \in G$ and an exponent

$$\alpha \in \{1, \ldots, \text{ord}(\phi_h)\}.$$

The secret key is $sk = \alpha$, and the public key is comprised of ϕ_h and ϕ_h^α, represented as

$$pk = (\phi_h(g_1), \ldots, \phi_h(g_n), \phi_{h^\alpha}(g_1), \ldots, \phi_{h^\alpha}(g_n)).$$

Encryption $\mathcal{E}$: to encrypt a message $m \in G$, represented as a word $m = w(g_1, \ldots, g_n)$ in the given generators, under public key $(\phi_h(g_1), \ldots, \phi_h(g_n), \phi_{h^\alpha}(g_1), \ldots, \phi_{h^\alpha}(g_n))$, choose an exponent $\beta \in \{1, \ldots, \text{ord}(\phi_h)\}$ and output

$$c = \Big(\underbrace{\phi_h(g_1)^\beta, \ldots, \phi_h(g_n)^\beta}_{\phi_h^\beta}, \underbrace{w(\phi_{h^\alpha}(g_1)^\beta, \ldots, \phi_{h^\alpha}(g_n)^\beta)}_{\phi_h^{\alpha\beta}(m)}\Big).$$

Decryption $\mathcal{D}$: to decrypt $c = (\phi_h^\beta, \phi_h^{\alpha\beta}(m))$ under secret key $sk = \alpha$, express $\phi_h^{\alpha\beta}(m)$ as a word $w'(g_1, \ldots, g_n)$ in the given generators, compute a representation of $\phi^{-\alpha\beta}$ as

$$((\phi_h(g_1)^\beta)^{-\alpha}, \ldots, (\phi_h(g_n)^\beta)^{-\alpha})$$

and recover the plaintext as

$$m = w'((\phi_h(g_1)^\beta)^{-\alpha}, \ldots, (\phi_h(g_n)^\beta)^{-\alpha}).$$

FIGURE 4.7: MOR encryption scheme.

The careful reader probably noticed already that the description in Figure 4.7 is imprecise, and it is not specified how to choose the exponents α, β

and the element h. Before trying to establish provable guarantees, certainly more details had to be filled in here. The original proposal suggested that, unlike for the ElGamal scheme, the exponent β can remain the same when encrypting a new plaintext and to exploit this for fast encryption/decryption. However, it was later proven by C. Tobias [Tob02] that if the adversary has access to a plaintext-ciphertext pair corresponding to the same exponent β then he will easily decrypt any further encryption with β (so, almost trivially, the scheme is not secure in the sense of IND-CCA1). Moreover, it can be proven that partial information on a hidden plaintext can actually be used to decrypt it. Finding concrete parameters for MOR that are not vulnerable to attack turned out to be a non-trivial problem, and the cryptanalytic successes are not very encouraging. $\Box$

From the mathematical point of view the absence of a provable security reduction for a cryptographic scheme is rather unfortunate, because this also means that a successful attack may actually not provide any conceptual insight. Because of Proposition 4.1, for ElGamal we are in a better situation. While it might very well be possible to violate the IND-CPA security of the latter, Proposition 4.1 ensures us that an attack would translate in a solution for the decisional Diffie–Hellman problem in the underlying group. In general, it is desirable that a provable security guarantee for a cryptographic scheme is reduced to a "cryptography-free" mathematical problem which is well studied and believed to be hard.

4.3   One-way functions and random oracles

A fundamental assumption for many cryptographic constructions, including public-key encryption, is the existence of one-way functions. Intuitively, these are efficiently computable functions with the property that finding a preimage of a "random image" is hard. To formalize this notion, the following approach can be used:

DEFINITION 4.8 [One-way function]   *Let $f : \{0,1\}^* \longrightarrow \{0,1\}^*$ be a probabilistic polynomial time computable function. We say that f is one-way if for all probabilistic polynomial time algorithms $\mathcal{A}$ the probability*

$$\Pr[f(\mathcal{A}(1^\ell, f(x))) = f(x) : x \leftarrow \{0,1\}^\ell]$$

is negligible. At this, the probability is taken over the uniform random choice of the ℓ-bit string x and over the internal coin tosses of $\mathcal{A}$.

If a one-way function f defines for each $\ell \in \mathbb{N}$ a bijection of $\{0,1\}^\ell$ into itself, we refer to f as a one-way permutation.

REMARK 4.5   Unless otherwise stated, we will talk about the one-wayness property of a function considering the input values x as being chosen uniformly at random, but note that the above definition generalizes naturally. The same observation applies to other definitions in this section.   □

Definition 4.8 characterizes a one-way function as a single function where finding preimages becomes increasingly harder when being presented with evaluations at inputs of increasing length. If we try to formalize problems like computing a discrete logarithm or decomposing a natural number into prime factors, this definition is not particularly convenient to work with. The functions we want to work with may be defined on a finite domain that depends on the security parameter and this domain is not necessarily defined as a certain subset of $\{0,1\}^*$, but as a more complex algebraic structure. The subsequent definition offers a formalization of the one-wayness property that takes this into account:

DEFINITION 4.9 [One-way function family]   *Let I be an index set. For each $i \in I$, let $f_i : D_i \longrightarrow R_i$ be a function such that the sets D_i and R_i are finite. The collection of functions $\{f_i : D_i \longrightarrow R_i\}_{i \in I}$ is a one-way function family, if there exist probabilistic polynomial time algorithms to*

- *select on input 1^ℓ an index $i \in I$,*

- *sample on input $i \in I$ a value $x \in D_i$ uniformly at random,*

- *compute on input i and $x \in D_i$ the value $f_i(x)$,*

in such a way that the following condition holds: for all probabilistic polynomial time algorithms $\mathcal{A}$ the probability

$$\Pr[f_i(\mathcal{A}(f_i(x), i, 1^\ell)) = f_i(x))]$$

is negligible, where the probability is taken over the coin tosses involved in the selection of i and x and the internal coin tosses of $\mathcal{A}$. A one-way permutation family is a one-way function family where each f_i is a permutation.

Example 4.6
Let $I = \mathbb{N}$ be the natural numbers. On input a security parameter 1^ℓ we choose the index $\ell \in I$ to satisfy the first requirement in the above definition. For any index $i \in I$ we define D_i to be the set of prime numbers (p, q) such that $p \neq q$ and $p, q \in \{2^i + 1, 2^i + 3, \ldots, 2^{i+1} - 1\}$. The set R_i is defined as $\{1, \ldots, 2^{i+2}\}$. Assuming the availability of a suitable sampling algorithm for the sets D_i, the multiplication functions $f_i : D_i \longmapsto R_i, (p, q) \mapsto p \cdot q$ form a candidate for a one-way function family which captures the idea that factoring a sufficiently large RSA modulus is not feasible.   □

It is most interesting for cryptographic purposes to have at hand seemingly hard computational problems that may however be efficiently solved with some (secret) additional information. The *trapdoor* property of one-way function families captures exactly this situation. More precisely, a one-way permutation family has a trapdoor if for each f_i there exists some value t_i allowing to compute efficiently the (unique) preimage of any given value in R_i. More formally, we have the following definition.

DEFINITION 4.10 [Trapdoor one-way function family]
Let $\mathbf{F} = \{f_i : D_i \to R_i\}_{i \in I}$ *be a one-way function family, and for each* $i \in I$ *let* T_i *be a finite set. We say that* $\mathbf{F}$ *is a* trapdoor one-way function family *if there exist probabilistic polynomial time algorithms for the following tasks:*

- *on input* 1^ℓ *output a pair* $(i, t_i) \in I \times T_i$,

- *on input* $(i, t_i, f_i(x))$, *for any* $i \in I$ *and* $x \in D_i$ *output* $x' \in D_i$ *with* $f_i(x') = f_i(x)$.

If each f_i *in a trapdoor one-way function family is a permutation, we call* $\mathbf{F}$ *a trapdoor one-way permutation family.*

As an example, let us define a candidate for a trapdoor one-way permutation family that describes the hardness of inverting basic RSA encryption.

Example 4.7
Denote by I the set of all tuples $(i, p, q, e) \in \mathbb{N}^4$ such that $p \neq q$ are prime, $p, q \in \{2^i + 1, 2^i + 3, \dots, 2^{i+1} - 1\}$ and $\gcd(e, \varphi(n)) = 1$. Assume we have a suitable algorithm which on input 1^ℓ samples a tuple of the form $(\ell, \cdot, \cdot, \cdot) \in I$ uniformly at random. For an index (i, p, q, e) we define D_i and R_i to be the group of units $\mathbb{Z}_{p \cdot q}^*$, and T_i is the singleton set containing the smallest positive integer d such that $d \cdot e - 1$ is divisible by $(p - 1) \cdot (q - 1)$. $\quad\square$

To simplify the presentation, we sometimes speak of a (trapdoor) one-way function f_ℓ where ℓ is the security parameter, suppressing the execution of the algorithm which derives the index i from 1^ℓ. While this should not cause any confusion, one should keep in mind that for a given security parameter ℓ the selection of a corresponding index i (and in case of a trapdoor one-way function family also the trapdoor) is in general randomized. In fact, it is easy to see that the derivation of a trapdoor from the security parameter cannot be deterministic.

Unfortunately, the construction of efficient and secure public-key encryption schemes turns out to be rather challenging, if we want to base our proof of security on a mathematical assumption like the decisional Diffie–Hellman assumption or the availability of a trapdoor one-way function family alone.

Known efficient constructions often make use of a cryptographic hash function H for which no formal security proof under the assumption in question is available, but when replacing H with an idealization known as *random oracle*, then a security proof becomes possible. Therefore we conclude this section with a brief introduction to the so-called *random oracle model* in which many of the constructions introduced in the sequel are defined. While "pathological" examples have shown that this model has clear limitations, these examples seem of meager practical concern, and random oracles have become a common tool in the design of cryptographic protocols. If a result involves neither random oracles nor any other idealizing assumptions, it is common to say that a result holds in the *standard model*.

Cryptographic hash functions are essential building blocks of many cryptographic constructions. Various flavors of *collision resistance*, capturing the hardness of finding different inputs providing the same hash value, are often the key argument in proofs. Moreover, many security arguments actually call for the stronger assumption that values output by a hash function look *random* to those ignoring the corresponding preimage. With the aim of providing a formal scenario capturing the use of cryptographic hash functions in practice, Bellare and Rogaway provided in [BR93] an *"explicitly articulated paradigm,"* the *random oracle model*, and applied it to derive efficient cryptographic schemes that could be proven secure in this model. For the case of encryption, their original proposal would later lead to a nowadays widely known RSA-based encryption method: RSA-OAEP, which we will present in Section 4.5. It addresses shortcomings in the basic version of RSA encryption we have seen in Section 3.1.2 (cf. Example 4.1).

The random oracle model formalizes a hash function as an oracle that produces a truly random value for each new query. Further, as cryptographic hash functions are deterministic, if the same input is queried to a random oracle repeatedly, always the same random value is returned. Commonly the set from which this output of the random oracle is selected uniformly at random increases exponentially in the security parameter ℓ—a typical choice are bitstrings of length ℓ or a cyclic group with at least 2^ℓ elements. A random oracle can thus be seen as a probabilistic machine, which on input a new value v outputs an element $H(v)$ chosen uniformly at random from the fixed finite image set. This machine stores all encountered input-output pairs $(v, H(v))$, so that repeated queries for the same value can be answered consistently.

REMARK 4.6 It is not uncommon to have a random oracle H with an infinite set of possible inputs (each of which has finite length), e. g., $\{0,1\}^*$. This provides no problem, as to answer a query, the probabilistic machine we just described only needs to sample uniformly at random in the finite output domain to answer a query. $\Box$

With the terminology established, we are now ready to describe the general encryption framework put forward in [BR93] and give a detailed security proof.

4.4 The general Bellare–Rogaway construction

For a fixed security parameter ℓ, suppose we choose the message space $\{0,1\}^\ell$ for our plaintexts together with a function $f_\ell : \{0,1\}^\ell \longrightarrow \{0,1\}^\ell$, chosen from a trapdoor one-way permutation family. Moreover, consider two cryptographic hash functions

$$G_\ell : \{0,1\}^\ell \longrightarrow \{0,1\}^\ell \text{ and } H_\ell : \{0,1\}^{2\ell} \longrightarrow \{0,1\}^{\ell_1},$$

where ℓ_1 is polynomial in ℓ (for asymptotic arguments, we may assume ℓ_1 grows roughly as ℓ). For the sake of readability, in the sequel we drop the subscript ℓ if there is no risk of confusion. We combine the random oracles G and H with the trapdoor one-way permutation family as shown in Figure 4.8 to form a public-key encryption scheme, where $\oplus$ represents component-wise addition modulo 2 (XOR).

$\mathcal{K}$: on input 1^ℓ outputs a description of a trapdoor one-way permutation f as public key pk and trapdoor information for evaluating the inverse f^{-1} of f as secret key sk.

$\mathcal{E}$: on input (pk, m), where $m \in \{0,1\}^\ell$, selects a value $r \leftarrow \{0,1\}^\ell$, computes

$$x = f(r), \ y = m \oplus G(r), \text{ and } z = H(m\|r),$$

and returns the ciphertext $c = x\|y\|z$.

$\mathcal{D}$: on input (sk, c) parses c as $c = x\|y\|z$ and computes

$$r = f^{-1}(x) \text{ and } m = y \oplus G(r).$$

If $z = H(m\|r)$ the algorithm returns m, otherwise, it returns $\perp$.

FIGURE 4.8: General Bellare–Rogaway construction.

We want to show that this scheme provides, in the random oracle model, security in the sense of IND-CCA. Namely, the proof assumes that G and H behave like random oracles, in the way we discussed in Section 4.3. The proof will be carried out using the (by now) standard *game hopping* technique [Sho04]. At this, the adversary is confronted with a sequence of challenge games, starting from the real game modeling the concrete attack and concluding with one in which it is clear that his probability of success is negligible. We will denote by P_i the probability of success of the adversary in the ith game. The result is then derived by bounding the difference of the probabilities of success in consecutive games. This proof is essentially taken from [Poi05], where more details on the reduction can be found.

THEOREM 4.1 [Security of Bellare–Rogaway construction]

The public-key encryption scheme using the general Bellare–Rogaway construction in Figure 4.8 is secure in the sense of IND-CCA in the random oracle model, under the assumption that the function f is chosen from a trapdoor one-way permutation family with domain $\{0,1\}^{\ell}$.

PROOF   Let us start by depicting the situation faced by the adversary when trying to defeat the IND-CCA security of the scheme: our probabilistic polynomial time adversary $\mathcal{A}$ will, after having received the public key to be attacked, have to choose two ℓ-bit messages m_0 and m_1. Then he will receive a $2\ell + \ell_1$ bit string, c^*, which is the encryption of one of the two chosen plaintexts, namely, there exists a random value r^* such that the challenge ciphertext is constructed as $c^* = f(r^*) \| m_b \oplus G(r^*) \| H(m_b \| r^*)$ with $b \in \{0,1\}$. The adversary's goal is to find out whether b equals 0 or 1. For this purpose, $\mathcal{A}$ has access to all public information, the random oracles G and H as well as to a

[Decryption oracle $\mathcal{O}_{\mathcal{D}}$]: on input a bit string $x\|y\|z$, with $x, y \in \{0,1\}^{\ell}$ and $z \in \{0,1\}^{\ell_1}$ it

1. computes $r = f^{-1}(x)$ and $m = y \oplus G(r)$,
2. checks whether $z = H(m\|r)$ and outputs m if this is the case and $\perp$ otherwise.

In the sequel, we will denote by $\texttt{GoodQuery}_i$ the event that in Game no. i any of the values involved in the construction of the challenge ciphertext were queried to the corresponding random oracles—i.e., either r^* was queried to G or $m_b\|r^*$ was queried to H.

Game G_0: This is exactly the "real" game modeling $\mathcal{A}$'s attack, thus, we have to prove that the absolute value of $P_0 - 1/2$ is upper bounded by some negligible function.

Game G_1: Let us assume that, instead of allowing $\mathcal{A}$ access to the random oracles G and H, we give him access to two simulators behaving as follows:

[Decryption oracle $\mathcal{O}_D$] : on input a bitstring $x\|y\|z$ with $x \in \{0,1\}^\ell, y \in \{0,1\}^\ell$, and $z \in \{0,1\}^{\ell_1}$ it

1. computes $r = f^{-1}(x)$ and $m = y \oplus G(r)$,

2. checks whether $z = H(m\|r)$ and outputs m if this is the case and $\perp$ otherwise.

[Challenge construction] On input m_0 and m_1,

1. selects $b \leftarrow \{0,1\}$ and $r^* \leftarrow \{0,1\}^\ell$,

2. computes

$$x = f(r^*), \ y = m_b \oplus G(r^*), \text{ and } z = H(m_b\|r^*),$$

3. outputs the challenge ciphertext $c^* = x\|y\|z$.

FIGURE 4.9: Real IND-CCA game.

[Simulator of Random oracle G] : on input a new value r, it selects $g \in \{0,1\}^\ell$ uniformly at random and stores (r, g). If G is then queried again on r, the output will be g.

[Simulator of Random oracle H] : on input a new value v, it selects $h \in \{0,1\}^{\ell_1}$ uniformly at random and stores (v, h). If H is then queried again on v, the output will be h.

The random oracle assumption actually states that these two games are identical, and so $|P_1 - P_0| = 0$.

Game G_2: At this, the adversary gets a challenge bitstring $x\|y\|z$ which is not fully constructed by encrypting faithfully the corresponding plaintext: x and y are constructed as in Figure 4.8, but z is just an element selected uniformly at random from $\{0,1\}^{\ell_1}$. Indeed, the difference between this and the "real" game may only be noticed if at some point the random oracle H is queried on the input with which z should have been constructed; namely, on $m_b\|r^*$. Therefore we have

$$|P_2 - P_1| \le \Pr[\texttt{GoodQuery}_2].$$

Game G_3: Now we modify the behavior of the decryption oracle $\mathcal{O}_D$, in that it rejects any input ciphertext for which the corresponding value $m\|r$ has not been queried to H. This game is thus identical to the previous one, except for the case in which the adversary has made a correct guess for a z-value, which happens with probability at most $q_{\mathcal{O}_D}/2^{\ell_1}$, where $q_{\mathcal{O}_D}$ denotes the (polynomial) number of queries to the decryption oracle made by the adversary.

Thus, now
$$|P_3 - P_2| \leq \frac{q \mathcal{O}_\mathcal{D}}{2^{\ell_1}}.$$

And, for the same reason,

$$|\Pr[\mathtt{GoodQuery}_3] - \Pr[\mathtt{GoodQuery}_2]| \leq \frac{q \mathcal{O}_\mathcal{D}}{2^{\ell_1}}.$$

Game G_4: In this game, during the challenge construction phase we deviate further from the scheme description, in that the value g is chosen uniformly at random from $\{0,1\}^\ell$, instead of being output by the random oracle G. This game cannot be distinguished from the previous one, except one of the following two situations occurs:

- at some point the random value r^* used for constructing the challenge ciphertext has been queried to the random oracle G by the adversary, which happens with probability at most $\Pr[\mathtt{GoodQuery}_4]$,

- the random value r^* used for constructing the challenge ciphertext is involved—even if not explicitly derived by the adversary—in the appropriate way in a certain input of the decryption oracle, but then it will subsequently be queried to G, and so it falls again under $\Pr[\mathtt{GoodQuery}_4]$.

Note also that this does not influence the probability of success of $\mathcal{A}$, as $m_b \oplus g$ with g random gives nothing more than $m_b \oplus G(r^*)$ to the adversary, and thus $P_3 = P_4$. Moreover, $\Pr[\mathtt{GoodQuery}_4] = \Pr[\mathtt{GoodQuery}_3]$.

Now, it is easy to argue that actually P_4 is exactly $1/2$, as the adversary is presented with a challenge ciphertext which is masked by a random value g that does not appear anywhere else.

Game G_5: Now we go one step further in modifying the challenge construction; namely, the first element in the challenge ciphertext, x, will now be selected uniformly at random from $\{0,1\}^\ell$. As f is a permutation, note that this change results in an experiment that is perfectly indistinguishable from G_4.

Game G_6: At this, the simulation of the decryption oracle is changed again: now it will reject if it gets an input involving an r-value that was never queried to G. Namely, for each input $(x\|y\|z)$ it searches through all values queried to G, evaluates f in them and rejects the ciphertext if x is not among these images.

This simulation differs from the previous one if for a given input to the decryption oracle $(x\|y\|z)$ the corresponding $m\|r$ was queried to H but r was never queried to G. As $G(r)$ would in this case be unpredictable and so is m (which is fully determined by y and r) the probability of such an H-query is bounded by $q_H/2^\ell$.

$$| \Pr[\texttt{GoodQuery}_6] - \Pr[\texttt{GoodQuery}_5]| \leq \frac{q_H}{2^\ell},$$

where q_H denotes the number of queries made to the random oracle H by the adversary.

Also, it is easy to argue that the event $\texttt{GoodQuery}_6$ only takes place if a preimage of x by f has been found, which only happens with negligible probability in ℓ.

Putting all the above together, we conclude P_1 must be negligible in ℓ, which yields the desired result.

$\square$

Game	Challenge construction	Decryption oracle
G_2	choose $b \leftarrow \{0,1\}, r \leftarrow \{0,1\}^\ell$, and $h \leftarrow \{0,1\}^{\ell_1}$, define $x = f(r), y = m_b \oplus G(r)$ and $z = h$	as in G_1
G_3	as in G_2	if $m\|r$ not queried to H, output $\bot$
G_4	choose $b \leftarrow \{0,1\}, r \leftarrow \{0,1\}^\ell$, $g \leftarrow \{0,1\}^\ell$ and $h \leftarrow \{0,1\}^\ell$, define $x = f(r), y = m_b \oplus g$ and $z = h$	as in G_3
G_5	choose $b \leftarrow \{0,1\}, a \leftarrow \{0,1\}^\ell$, $g \leftarrow \{0,1\}^\ell$ and $h \leftarrow \{0,1\}^\ell$, define $x = a, y = m_b \oplus g$ and $z = h$	as in G_4
G_6	as in G_5	if no r' queried to G s.t. $f(r') = g$, output $\bot$

FIGURE 4.10: Bellare–Rogaway security proof in a nutshell.

A main problem of the above construction is the high overhead; namely, ciphertexts of ℓ bit messages are of size $2 \cdot \ell + \ell_1$. The *Optimal Asymmetric Encryption Padding (OAEP)* is one attempt to come up with a more compact construction for a public-key encryption scheme, which offers IND-CCA security. OAEP was introduced by Bellare and Rogaway in [BR94b], and combining it with RSA as one-way permutation yields a much stronger scheme than the basic version of RSA we discussed at the beginning of this book.

4.5   IND-CCA security with an Abelian group: RSA-OAEP

The tools needed for an OAEP instantiation, with security parameter ℓ, are a trapdoor one-way permutation $f_\ell : \{0,1\}^\ell \to \{0,1\}^\ell$ and two cryptographic hash functions (random oracles)

$$G_\ell : \{0,1\}^{\ell_0} \to \{0,1\}^{\ell-\ell_0} \text{ and } H_\ell : \{0,1\}^{\ell-\ell_0} \to \{0,1\}^{\ell_0},$$

for some $\ell_0 < \ell$. We shall also fix a positive integer ℓ_1 and define $n = \ell + \ell_0 + \ell_1$. Thus, we assume fixed a trapdoor one-way permutation family $\{f_\ell\}_{\ell \in \mathbb{N}}$ along with two corresponding families of cryptographic hash functions $\{H_\ell\}_{\ell \in \mathbb{N}}$ and $\{G_\ell\}_{\ell \in \mathbb{N}}$, which we will model as random oracles. In what follows we drop the subscript ℓ if there is no ambiguity. Furthermore, if messages are of length n bits we fix $\ell_1 = \ell - n - \ell_0$.

Given the above, a public-key encryption scheme $\mathcal{P} = (\mathcal{K}, \mathcal{E}, \mathcal{D})$ can be constructed as described in Figure 4.11.

$\mathcal{K}$: on input 1^ℓ outputs a description of the function f (the public key pk) and of its inverse f^{-1} (the secret key sk).

$\mathcal{E}$: on input (pk, m) selects $r \leftarrow \{0,1\}^{\ell_0}$, computes

$$s = (m||0^{\ell_1}) \oplus G(r) \text{ and } t = r \oplus H(s),$$

and outputs the ciphertext $c = f(s, t)$.

$\mathcal{D}$: on input (sk, c) computes

$$(s, t) = f^{-1}(c), \quad r = t \oplus H(s) \text{ and } M = s \oplus G(r).$$

Now, if the last ℓ_1 bits of M are zero, the algorithm returns the string m consisting of the first n bits of M. Otherwise, it returns $\perp$.

FIGURE 4.11:   Public-key encryption using OAEP.

The above scheme can be proven secure in the sense of IND-CCA1, and for some time it was widely believed to actually achieve IND-CCA security. However, Victor Shoup provided in [Sho01, Sho02] evidence that to that aim requiring f to be a one-way permutation is not enough. Later, Eiichiro Fujisaki, Tatsuaki Okamoto, David Pointcheval, and Jacques Stern (see [FOPS01, FOPS04]) came up with a proof of IND-CCA security at the price of imposing a stricter condition on the function (family) f:

DEFINITION 4.11 [Partial-domain one-wayness]
Suppose that $\{f_i : D_i^1 \times D_i^2 \to R_i\}_{i \in I}$ is a collection of permutations. We refer to this collection as a partial domain one-way permutation family *provided that there exist probabilistic polynomial time algorithms to:*

- *select on input 1^ℓ an index $i \in I$,*

- *select on input $i \in I$ a value $(x^1, x^2) \in D_i^1 \times D_i^2$,*

- *compute on input i and (x^1, x^2) the value $f_i(x^1, x^2)$*

in such a way that, for all probabilistic polynomial time algorithms $\mathcal{A}$, there exists a negligible function ε such that

$$\Pr[\mathcal{A}(f_i(x^1, x^2), i, 1^\ell) = x^1] \le \varepsilon(\ell),$$

where the probability is taken over the randomness used by $\mathcal{A}$ and the coin tosses involved in the selection of i and $(x^1, x^2) \in D_i^1 \times D_i^2$.

Comparing the above definition with the definition of "complete" one-wayness used earlier (Definition 4.9), we see that partial-domain one-wayness simplifies the adversary $\mathcal{A}$'s task in the sense that $\mathcal{A}$ now has to recover only a part of the preimage—for a permutation this can be regarded as a simplification, as here the preimage is unique. Now, under the assumption that f is taken from a partial domain one-way permutation family, the following exact security result has been established in [FOPS01]; we do not include the proof of this result here:

THEOREM 4.2 [Exact security of OAEP]
Let $\mathcal{A}$ be a probabilistic polynomial time adversary winning the IND-CCA game with probability $\frac{1}{2} + \varepsilon$ against the OAEP scheme described in Figure 4.11. Assume $\mathcal{A}$ does so in time τ making $q_{\mathcal{O}_D}$, q_G, q_H queries to the decryption oracle and the hash functions G and H, respectively.
  Then, the probability $\Pr[\mathcal{A}(s, t) = s]$ that $\mathcal{A}$ succeeds in violating the partial-domain one-wayness of f in time τ', is lower-bounded by

$$\frac{1}{q_H} \cdot \left(\frac{\varepsilon}{2} - \frac{2q_d \cdot q_G + q_{\mathcal{O}_D} + q_G}{2^{\ell_0}} - \frac{2q_d}{2^{\ell_1}} \right),$$

where $\tau' \le \tau + q_G \cdot q_H \cdot (T_f + \mathcal{O}(1))$. At this, T_f denotes the time complexity of f.

REMARK 4.7   The above theorem looks different than the security result we have seen in Theorem 4.1 for the Bellare–Rogaway construction, but it is also a proof of IND-CCA security: if f is partial-domain one-way, then ε is a negligible function, and Theorem 4.2 in particular states that OAEP-base d encryption is then IND-CCA secure (in the random oracle model). However, the bound provided in the theorem gives us more than this asymptotic statement: we have a quantitative assurance about the probability with which we can violate our hardness assumption, in terms of resources used by $\mathcal{A}$. This type of result is thus more helpful when aiming at a concrete instantiation of the scheme at hand.   ▯

Instantiating OAEP taking f as an RSA function (as defined in Section 3.1.2) is somewhat a fortunate case, as the partial-domain one-wayness of RSA is equivalent to that of the whole RSA problem. This is due to a property of the RSA function, called *random self-reducibility*; namely, it is possible to compute a concrete preimage of a given value, provided that certain (randomly selected in some sense) instances of this same problem can be solved.

4.6   One-way functions from non-Abelian groups?

Constructions like OAEP or the general Bellare–Rogaway construction offer a mathematically luring approach to designing a new public-key encryption scheme: if we can use a mathematical hardness assumption to derive a (family) of functions with a suitable trapdoor one-wayness property, these constructions take care of the "cryptographic technicalities" like chosen-ciphertext attacks.

With the multiplicative group $\mathbb{Z}_n^*$ used in RSA being a rather special (and from a group-theoretic point of view not particularly complicated) group, it is tempting to take a look at other (non-Abelian) groups to derive trapdoor one-way functions that can be combined with, e. g., OAEP. Interestingly, this task turns out to be quite challenging, and so far no non-Abelian construction has been identified that is widely accepted within the cryptographic research community. In this section we want to discuss one approach, which builds on the public-key scheme known as MST_1—proposed by Spyros Magliveras, Douglas Stinson, and Tran van Trung in [MSvT02]—and which naturally yields to a conjecture about the existence of certain types of factorizations of finite groups.

DEFINITION 4.12 [Logarithmic signature of a finite group]
Let G be a finite group and denote by $\alpha = [\alpha_1, \ldots, \alpha_s]$ a sequence of length $s \in \mathbb{N}_0$, where each block $\alpha_i = [\alpha_{i0}, \ldots, \alpha_{ir_i-1}] \in G^{r_i}$ is a sequence of $r_i \in \mathbb{N}_0$ group elements.

If every element $g \in G$ can be expressed as a product

$$g = \alpha_{1j_1} \cdot \cdots \cdot \alpha_{sj_s} \qquad (4.4)$$

with $\alpha_{ij_i} \in \alpha_i$ ($1 \leq i \leq s$), then we refer to α as a cover *of G. Further, if α is a cover of G and the representation (4.4) is unique for all $g \in G$, then α is called a* logarithmic signature *of G.*

REMARK 4.8 The term "logarithmic signature" is motivated by taking the elements in α for a compact representation of the group G—a logarithmic signature comprised of s blocks of size 2 describes a group of size 2^s. ⬚

From the definition it is clear that each finite group G has at least one trivial logarithmic signature—we can always set $\alpha = [\alpha_1]$ with α_1 containing all the elements of the group G in some order. Actually, according to Definition 4.12 changing the order in which we list the elements of G in α_1 would yield a different logarithmic signature, and so we have for each group G at least $\mathrm{ord}(G)!$ (trivial) logarithmic signatures. It is also obvious that the product of all block lengths occurring in a logarithmic signature of a finite group G must be equal to the order of G, i.e., with the notation from Definition 4.12 we have

$$r_1 \cdot \cdots \cdot r_s = \mathrm{ord}(G).$$

Consequently, if we exclude blocks of size 1, for finite groups of prime order, trivial logarithmic signatures consisting of a single block are the only ones that exist. On the other hand, if the group G has a non-trivial subgroup $G_1 < G$, then we can derive logarithmic signatures for G in a simple way: if the sequence α_2 is a complete set of right-coset representatives of G_1 in G (i.e., a right-transversal for G_1 in G) and we store in α_1 the elements of the subgroup G_1 in some order, then $[\alpha_1, \alpha_2]$ is a logarithmic signature for G. Analogously, if we choose β_1 to be a left-transversal of G_1 in G and store in β_2 the elements of G_1 in some order, then $[\beta_1, \beta_2]$ is a logarithmic signature for G. Iterating this process, from a subgroup chain

$$G = G_0 > G_1 > \cdots > G_{s-1}$$

we can derive a logarithmic signature for G which consists of s blocks. For the obvious reason, logarithmic signatures of this type are often referred to as *transversal*; note that G_{s-1} can be seen as a transversal as well, namely as a transversal of the 1-element subgroup in G_{s-1}.

Example 4.8

Consider the subgroup chain $S_5 > A_5 > A_4 > V_4 > C_2$ inside the symmetric group on 5 points, where

$$S_5 = \langle (1,2,3,4,5), (1,2) \rangle$$
$$A_5 = \langle (1,2)(3,4), (2,4,5) \rangle$$
$$A_4 = \langle (1,5,3), (1,2)(3,5), (1,5)(2,3) \rangle$$
$$V_4 = \langle (1,2)(3,5), (1,5)(2,3) \rangle$$
$$C_2 = \langle (1,2)(3,5) \rangle.$$

We can choose, e. g., the left-transversals $[\mathrm{id}, (1,2)]$ of A_5 in S_5, $[\mathrm{id}, (1,5)(2,3)]$ of C_2 in V_4, and the right-transversals $[\mathrm{id}, (1,2)(3,4), (2,4,5), (2,5,4), (1,2,4)]$ of A_4 in A_5, $[\mathrm{id}, (1,5,3), (1,3,5)]$ of V_4 in A_4. In this way we obtain the following logarithmic signature for S_5:

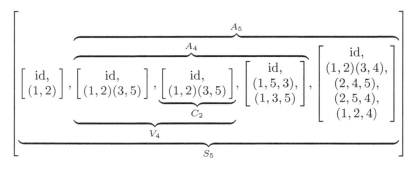

$\Box$

From logarithmic signatures for a finite group G of order n we can derive permutations of $\mathbb{Z}_n$. Suppose $\alpha = [\alpha_1, \dots, \alpha_s]$ is a logarithmic signature for G with $\alpha_i = [\alpha_{i0}, \dots, \alpha_{ir_i - 1}]$, then we have a natural bijection

$$
\begin{aligned}
f_\alpha : \mathbb{Z}_{r_1} \times \cdots \times \mathbb{Z}_{r_s} &\longrightarrow G \\
(i_1, \dots, i_s) &\longmapsto \alpha_{1i_1} \cdots \cdots \alpha_{si_s}.
\end{aligned}
$$

Moreover, as $n = r_1 \cdots r_s$, a mixed radix representation of integers gives us a bijection

$$
\begin{aligned}
\mu_{(r_1, \dots, r_s)} : \mathbb{Z}_{r_1} \times \cdots \times \mathbb{Z}_{r_s} &\longrightarrow \mathbb{Z}_n \\
(i_1, \dots, i_s) &\longmapsto i_1 + i_2 \cdot r_1 + i_3 \cdot r_1 r_2 + \cdots + i_s \cdot \prod_{j=1}^{s-1} r_j.
\end{aligned}
$$

Thus, if α is a logarithmic signature with block lengths $r_1, \dots, r_s$ for the finite group G and β is a logarithmic signature for G with block lengths $r'_1, \dots, r'_{s'}$ then we can form the permutation

$$
\begin{aligned}
\pi_{\alpha, \beta} : \mathbb{Z}_n &\longrightarrow \mathbb{Z}_n \\
x &\longmapsto \mu_{(r'_1, \dots, r'_{s'})} \circ f_\beta^{-1} \circ f_\alpha \circ \mu_{(r_1, \dots, r_s)}^{-1}(x).
\end{aligned}
\tag{4.5}
$$

Both $\mu_{(.)}$ and $\mu_{(.)}^{-1}$ can be evaluated efficiently, and if we assume the group operations in G to be computable efficiently, then f_α can be evaluated efficiently, too. If, moreover, the logarithmic signature β can be chosen such that factoring along β can be done efficiently, then $\pi_{\alpha,\beta}$ can be evaluated efficiently and the task of evaluating $\pi_{\alpha,\beta}^{-1}$ translates into finding the unique factorization of a group element along the logarithmic signature α.

Can we use this to construct a one-way permutation (maybe even a trapdoor one-way permutation if efficiently factoring elements with respect to α can be conditioned on the availability of some trapdoor information)? While this approach is theoretically appealing, turning it into an efficient algorithm to generate (plausible candidates for) trapdoor one-way permutations turns out to be challenging. So far no concrete construction has been widely accepted by the cryptographic research community. Once the problem of finding such logarithmic signatures can be overcome, however, from the Bellare–Rogaway construction in Figure 4.8 we could obtain (in the random oracle model) an IND-CCA secure public-key encryption scheme.

Also, even if a logarithmic signature α fulfilling the security requirements is at hand, there are additional prerequisites we should not neglect when aiming at a practical implementation. Indeed, to publish a description of the function f_α (or of the permutation $\pi_{\alpha,\beta}$) it appears natural to provide a description of the logarithmic signature α. Unless α comes with some additional structure, the obvious way to write down α is to list the group elements in each block of α, and if α becomes part of the public key of an encryption scheme, one may be interested in keeping the public key as short as possible.

DEFINITION 4.13 [Length of a logarithmic signature]   *For a logarithmic signature $\alpha = [\alpha_1, \ldots, \alpha_s]$ with $\alpha_i \in G^{r_i}$ being of length r_i, we call $\mathrm{len}(\alpha) = r_1 + \cdots + r_s$ the length of α.*

If is not hard to see that the length of any logarithmic signature for a finite group G is lower-bounded by the sum of the prime divisors of $\mathrm{ord}(G)$.

PROPOSITION 4.2 [Minimal length of logarithmic signatures]
If G is a finite group and $\mathrm{ord}(G) = p_1 \cdot \cdots \cdot p_t$ the prime factorization of $\mathrm{ord}(G)$, then for each logarithmic signature α of G, we have

$$\mathrm{len}(\alpha) \geq p_1 + \cdots + p_t.$$

PROOF   If $\alpha = [\alpha_1, \ldots, \alpha_s]$ is a logarithmic signature for G with $\alpha_i \in G^{r_i}$ $(i = 1, \ldots, s)$, then $\mathrm{ord}(G) = r_1 \cdot \cdots \cdot r_s$. Decomposing each r_i into a product of prime numbers $r_i = p_{i,1} \cdot \cdots \cdot p_{i,t_i}$, the claim follows from the inequality $p_{i,1} \cdot \cdots \cdot p_{i,t_i} \geq p_{i,1} + \cdots + p_{i,t_i}$. ☐

The logarithmic signature in Example 4.8 meets this bound: we have $\text{ord}(S_5) = 5 \cdot 3 \cdot 2 \cdot 2 \cdot 2$, and the logarithmic signature given consists of $14 = 2 + 2 + 2 + 3 + 5$ elements. It can be proved that if a finite group without such a "minimal length logarithmic signature" exists, then the smallest such group has to be simple (see Exercise 33). Consequently, the classification of finite simple groups offers a roadmap to try to establish the existence of such a minimal logarithmic signature for all finite groups. For various simple groups the existence has been demonstrated, but at the time of writing this text, the general case is still open.

The above discussion already reveals that developing cryptographic tools from group theory is a rather involved task; not only strict security requirements have to be taken into account, but there are also implementation-level imperatives, which, if neglected, may lead to failure of a nice mathematical idea.

4.7 Summary and further reading

In this chapter we have

- introduced the basic formal definitions related to public-key encryption, some fundamental security notions for encryption schemes as well as one-way (trapdoor) function families.

- presented the random oracle model and a generic construction of a secure scheme achieving IND-CCA security in this model. This allowed us to see a first security proof using the game-hopping technique.

- discussed a method for developing IND-CCA secure schemes from the above construction using as a base a group-theoretic problem: factoring with respect to logarithmic signatures of finite groups. The problem of finding logarithmic signatures of minimal length has been presented (as an example of a nice group-theoretic question arising from a cryptographic context).

For a brief survey on provable security for public-key encryption and signature schemes, see [CCD+05, Chapter D] by D. Pointcheval. Game hopping proofs may seem a bit intricate for the inexperienced reader; V. Shoup's tutorial on this technique [Sho04] can be highly recommended. Another interesting (and rather controversial) writing is that of [KM07], where an informal exposition and critique of the provable security paradigm is presented.

For readers interested in knowing more about the random oracle model, besides looking at the seminal paper [BR93] in which it is introduced, it is indeed worth reading the work of Canetti et al. [CGH04]. In this paper, the limits of the model become explicit as the authors present a scheme that is secure in the random oracle model but can be proven insecure without using an argument that may imply any kind of weakness in the used hash functions. All in all, proofs in this model provide a sound assurance that the corresponding schemes can be used with security guarantees in practice.

A nice and accessible introduction to (number theoretical) cryptographic hardness assumptions can be found in [KL07, Chapter 7]. Exploring the implications of cryptographic assumptions is however (to some extent surprisingly) an active research area; see, for instance, the recent results from [EHK+13, JS13].

Basic terminology, (sometimes striking) examples, and very precise definitions of different types of one-way functions can be found in Golreich's *Foundations of Cryptology (Vol. I)* [Gol01, Chapter 2].

Prior to the attempt of using logarithmic signatures in public-key cryptography, these tools were at the core of the definition of a symmetric encryption scheme called PGM [MM92], which we will get back to in Section 7.3. A public-key encryption scheme using a logarithmic signature as a base would naturally aim at an instantiation of some sort of trapdoor one-way function family as we have discussed above [MSvT02]. There were however problems with the key generation procedure and finding a way to derive suitable logarithmic signatures to be used as key turned out to be rather troublesome (see [BSGVM05]). A follow up proposal along this line can be found in [LvTMW09]; see also the corresponding cryptanalysis in [BCM09, GVPdPTD10]. Some results on minimal length logarithmic signatures can be found in [GVRS03, LvT05] and in more recent papers focusing on sporadic [Hol04] and Lie type groups [SSM10].

4.8 Exercises

Exercise 24 *Discuss the security of two variants of the Bellare–Rogaway construction (Figure 4.8), derived by modifying the encryption algorithm as follows:*

[BR-α] $\mathcal{E}$, *on input* (pk, m), *where* $m \in \{0, 1\}^{\ell}$, *selects uniformly at random a value* r *in* X *and computes*

$$x = f(r), \ y = m \oplus G(r) \, and \, z = H(m);$$

[BR-β] $\mathcal{E}$, *on input* (pk, m), *where* $m \in \{0, 1\}^{\ell}$, *selects uniformly at random a value* r *in* $\{0, 1\}^{\ell}$ *and computes*

$$x = f(r), \ y = m \oplus G(r) \, and \, z = H(r).$$

Exercise 25 *The ElGamal encryption scheme may seem inefficient in the sense that encryption of a single group element yields a ciphertext consisting of two group elements.*

   (a) Show that there exists no IND-CPA secure public-key encryption scheme that works "in place" in the sense that the finite set of possible plaintexts has the same size as the set of possible ciphertexts.

   (b) Prove that a public-key encryption scheme with a deterministic encryption algorithm cannot be secure in the sense of IND-CPA.

Exercise 26 *In the definition of OW-CPA, we required the* advantage *of all probabilistic polynomial time adversaries $\mathcal{A}$ to be negligible. Suppose we would instead have imposed the* success *probability of all probabilistic polynomial time adversaries to be negligible. Show that with such a definition security in the sense of IND-CPA does not necessarily imply security in the sense of OW-CPA.*

Exercise 27 *Suppose we are given a public-key encryption scheme that is secure in the sense of IND-CPA and can encrypt one bit, i. e., the plaintext space is $\{0,1\}$. Explain how this scheme can be used to encrypt, for any fixed positive integer n, plaintexts in $\{0,1\}^n$ such that security in the sense of IND-CPA is ensured.*

   If the scheme to encrypt one bit is IND-CCA secure, does your construction to encrypt n-bit strings ensure security in the sense of IND-CCA?

Exercise 28 *Consider a ciphertext $c = (g^\beta, g^{\alpha \cdot \beta} \cdot m)$ that has been obtained by encrypting an unknown group element m under a public key $pk = (g, g^\alpha)$ with the ElGamal encryption scheme. Show how an adversary can derive an encryption c' of the inverse m^{-1} under pk.*

Exercise 29 *Explain how the hardness of a discrete logarithm problem can be formalized in terms of a one-way function family.*

Exercise 30 *Some authors formulate hardness assumptions or security requirements in a slightly different manner than we did above: Instead of requiring that for each probabilistic polynomial time algorithm $\mathcal{A}$ there exists a negligible function $n_\mathcal{A}(\ell)$ bounding the (suitably defined) advantage of $\mathcal{A}$, the existence of a single ($\mathcal{A}$-independent) negligible function $n(\ell)$ is imposed such that $n(\ell)$ bounds the advantage of* all *probabilistic time algorithms $\mathcal{A}$.*

   Give an example of a family of functions $\{n_i : \mathbb{N} \longrightarrow \mathbb{R}_{\geq 0}\}_i$ such that each function n_i of this family is negligible, but there exists no negligible function $n : \mathbb{N} \longrightarrow \mathbb{R}_{\geq 0}$ such that $n_i(\ell) \leq n(\ell)$ for all i for sufficiently large ℓ.

Exercise 31 *Describe an adversary $\mathcal{A}$ which shows that ElGamal encryption does not guarantee ciphertext indistinguishability under an adaptive chosen ciphertext attack.*

Exercise 32 *Let $G = \langle g \rangle$ be a cyclic group of order p^ℓ where p is a prime number and $\ell \in \mathbb{N}$ is a positive integer. Show that there is a logarithmic signature α for G which meets the length bound in Proposition 4.2 and such that knowing the factorization $h \in G$ along α is equivalent to knowing $\log_g(h)$.*

Exercise 33 *Suppose that G is a finite group and H a normal subgroup of G such that both H and G/H have a logarithmic signature meeting the length bound in Proposition 4.2. Prove that G has such a minimal length logarithmic signature as well. Conclude that the smallest group for which no minimal length logarithmic signature exists (if there is such a group) has to be simple.*

Chapter 5

Public-key encryption in the standard model

In Section 4.2 we have seen that security in the sense of IND-CCA is equivalent to security in the sense of NM-CCA, whereas for adversaries which are restricted to chosen plaintext attacks, non-malleability is not implied by ciphertext indistinguishability. Thus, if we aim at boosting the security of a public-key encryption scheme so that it withstands chosen ciphertext attacks, a plausible strategy is to append a "proof of integrity" during the encryption process. Namely, as part of the ciphertext a tag is included, that ensures the ciphertext was properly constructed and not produced by cleverly tampering with known ciphertexts. This "proof of integrity" may be derived using random oracles (as is the case in the Bellare–Rogaway construction from Figure 4.8), but as we will see in this chapter it is also possible to avoid their use.

Ronald Cramer and Victor Shoup presented in 1998 the first efficient public-key encryption scheme that could be proven IND-CCA secure in the standard model under a plausible number theoretic assumption [CS98]. A few years later, the same authors introduced in [CS02] a new paradigm for constructing IND-CCA secure encryption schemes using *universal hash proof systems* for languages. This, in particular, allowed for the construction of the aforementioned "proofs of integrity" without random oracles, and their 1998 construction is an instantiation of this paradigm. In this chapter we go through the main results in these two papers, and moreover discuss a way to adapt their ideas to work with non-Abelian groups.

5.1 The Cramer–Shoup encryption scheme from 1998

We start by formalizing a type of collision resistance, which one expects a cryptographic hash function as mentioned in Section 4.3 to satisfy. As for any fixed non-injective function f there is always an efficient algorithm $\mathcal{A}_f$ that can output a collision for f (cf. Exercise 56), instead of discussing collision resistance of a single function we look at a family of functions.

DEFINITION 5.1 [Collision-resistant hash functions]   *Let $L \subseteq \mathbb{N}$ and for each $\ell \in L$ consider $\mathcal{H}_\ell$, a finite collection of functions with range $\{0,1\}^\ell$ and having a common finite domain of definition D_ℓ. We say that the family $\mathcal{H} = \{\mathcal{H}_\ell\}_{\ell \in L}$ is* collision resistant *if any probabilistic polynomial time algorithm $\mathcal{A}$ has negligible probability (in ℓ) of, given a function H selected uniformly at random from $\mathcal{H}_\ell$, finding two values $x, y \in D^\ell$ such that*

$$H(x) = H(y) \text{ and } x \neq y.$$

A weaker requirement than collision resistance is *target collision resistance*, where an adversary has to commit in advance to one of the colliding inputs.

DEFINITION 5.2 [Target collision resistant hash functions]   *Let $L \subseteq \mathbb{N}$ and for each $\ell \in L$ consider $\mathcal{H}_\ell$, a finite collection of functions with range $\{0,1\}^\ell$ and having a common finite domain of definition D_ℓ. We say that the family $\mathcal{H} = \{\mathcal{H}_\ell\}_{\ell \in L}$ is* target collision resistant *if every probabilistic polynomial time algorithm $\mathcal{A}$ has negligible probability (in ℓ) in winning the following game: first $\mathcal{A}$ chooses $x \in D_\ell$, and thereafter is given a function H, chosen uniformly at random from $\mathcal{H}_\ell$. To win, $\mathcal{A}$ has to output $y \neq x$ such that $H(x) = H(y)$.*

REMARK 5.1   In the literature, target collision resistant hash functions are also called *universal one-way hash functions (OWHF)*. We prefer to use the above nomenclature to avoid confusion with the notion of *universal hashing*, which will be used extensively later in this chapter.   ⬜

The encryption scheme proposed by Cramer and Shoup at CRYPTO '98 assumes that a target collision resistant hash family is available and makes use of a prime order subgroup of the multiplicative group $\mathbb{F}_p^*$ of a finite prime field. Figure 5.1 describes the algorithms for key generation, encryption, and decryption.

Before pondering the security of this scheme, let us briefly argue that it is correct: indeed, if a ciphertext $c = (c_1, c_2, c_3, c_4)$ was properly formed, there exist $m, r \in G$ such that $c_1 = g_1^r, c_2 = g_2^r$ and $c_3 = h^r m$. Thence we have $c_1^{x_1} c_2^{x_2} = g_1^{r x_1} g_2^{r x_2} = c^r$ and also $c_1^{y_1} c_2^{y_2} = d^r$ and thus

$$c_4 = c^r d^{r\alpha} = c_1^{x_1} c_2^{x_2} (c_1^{y_1} c_2^{y_2})^{H(c_1, c_2, c_3)}.$$

So the test performed by the decryption algorithm will succeed and the output will be the encrypted message m.

To understand the security of Cramer and Shoup's construction, let us take a closer look at how the ciphertext $c = (c_1, c_2, c_3, c_4)$ of a message m is constructed: the message m itself is hidden in c_3, masked by a (uniformly at random) chosen group element h^r. The necessary information to remove this

---

$\mathcal{K}$: on input 1^ℓ, this algorithm

- outputs a description of a cyclic subgroup $G \le \mathbb{F}_p^*$ of prime order q (in particular $q|(p-1)$);

- outputs a description of a function $H : G^3 \longrightarrow \mathbb{Z}_q$ chosen uniformly at random from a suitable target collision resistant hash family;

- selects elements $g_1, g_2 \leftarrow G$;

- selects elements $x_1, x_2, y_1, y_2, z \leftarrow \{1, \ldots, q\}$;

- computes $c = g_1^{x_1} g_2^{x_2}$, $d = g_1^{y_1} g_2^{y_2}$, and $h = g_1^z$;

- outputs the public key $pk = (g_1, g_2, c, d, h)$ and the private key $sk = (x_1, x_2, y_1, y_2, z)$.

$\mathcal{E}$: on input pk as above and $m \in G$, this algorithm

- selects $r \leftarrow \{1, \ldots, q\}$,

- computes $u_1 = g_1^r, u_2 = g_2^r, e = h^r m, \alpha = H(u_1, u_2, e)$ and $v = c^r d^{r\alpha}$,

- outputs the ciphertext $c = (u_1, u_2, e, v)$.

$\mathcal{D}$: on input the secret key sk and a ciphertext $c = (c_1, c_2, c_3, c_4)$, this algorithm

- computes $\hat{\alpha} = H(c_1, c_2, c_3)$ and tests if $c_1^{x_1 + y_1 \hat{\alpha}} c_2^{x_2 + y_2 \hat{\alpha}} = c_4$;

- if the above check fails, the output is $\perp$, otherwise it is c_3 / c_1^z.

FIGURE 5.1: Cramer–Shoup public-key encryption from 1998.

mask is contained in c_1, as $c_1^z = g_1^{rz} = h^r$. Now, c_4 is a "proof of integrity," which allows the decryption algorithm to verify that the ciphertext indeed comes from a legitimate encryption process. This check can be done using the element g_2^r, which is actually encoded in c_2. Of course, for a more precise security argument it is crucial that no information about the mask h^r can be retrieved from the public values. This argument is ultimately based on the decisional Diffie–Hellman assumption (with uniform random selection of the group generator), but for the proof we use a formulation of the needed assumption which at first glance looks different than the one from Assumption 4.1. Proving the equivalence of the two formulations is the subject of Exercise 35.

Solving the decisional Diffie–Hellman problem can be seen as the task of identifying an efficient algorithm that can distinguish between two particular probability distributions: tuples of the form $(g, g^\alpha, g^\beta, g^\gamma)$ with uniformly at random chosen $\alpha, \beta, \gamma \in \{1, \ldots, \mathrm{ord}(g)\}$ and tuples of the form $(g, g^\alpha, g^\beta, g^{\alpha\beta})$ with uniformly at random chosen $\alpha, \beta \in \{1, \ldots, \mathrm{ord}(g)\}$. The assumption we make on the group (family) G used in the Cramer–Shoup scheme from 1998 is of the same nature, but uses slightly different tuples.

Assumption 5.1 (Cramer–Shoup decisional assumption) *Let G be a finite cyclic group. Then there exists no probabilistic polynomial time algorithm which can distinguish the following two probability distributions with more than negligible probability:*

- *the distribution of 4-tuples (g_1, g_2, g_3, g_4), where each g_i $(i = 1, \ldots, 4)$ is chosen uniformly at random from G,*

- *the distribution of 4-tuples (g_1, g_2, g_1^r, g_2^r), where g_1, g_2 are chosen uniformly at random from G, and r is chosen uniformly at random from $\{1, \ldots, \mathrm{ord}(G)\}$.*

Note that, if we believe the above assumption, the mask h^r used to hide the message m could be replaced, without a probabilistic polynomial time adversary noticing, by an element chosen uniformly at random from the group. Now, if the mask were actually distributed uniformly at random, then the message would be perfectly hidden, which yields the desired security in the sense of IND-CCA. The proof of the next theorem (taken from [CS98]) makes this idea more precise.

THEOREM 5.1 [Security of Cramer–Shoup '98 encryption]
Let us assume that, in the encryption scheme from Figure 5.1, the hash function H is chosen from a target collision resistant hash family. Then, provided that Assumption 5.1 holds for the underlying group (family) G, this encryption scheme is secure against adaptive chosen ciphertext attacks in the sense of indistinguishable encryptions (IND-CCA).

PROOF Let $\mathcal{A}$ be a probabilistic polynomial time adversary which succeeds in violating the IND-CCA security of the scheme, having access to a decryption oracle $\mathcal{O}_\mathcal{D}$ which will decrypt any query different from the challenge ciphertext. Let us see how we can use this adversary to construct a polynomial time distinguisher $\mathcal{D}$ for the two distributions from Assumption 5.1, namely:

- **D$_1$:** The distribution of 4-tuples (g_1, g_2, r_1, r_2), where g_1, g_2, r_1, r_2 are chosen uniformly at random from G,

- **D$_2$:** The distribution of 4-tuples (g_1, g_2, g_1^r, g_2^r), where g_1, g_2 are chosen uniformly at random from G, and r is chosen uniformly at random from $\mathbb{Z}_q$.

Our distinguisher $\mathcal{D}$ works as follows: on input a tuple (g_1, g_2, g_3, g_4), it will serve as challenger for $\mathcal{A}$ and interact several times with the latter, repeating the following strategy. First, $\mathcal{D}$

- selects values $x_1, x_2, y_1, y_2, z_1, z_2$ uniformly at random from $\mathbb{Z}_q$ (these will constitute his private key),

- computes $c = g_1^{x_1} g_2^{x_2}, d = g_1^{y_1} g_2^{y_2}$ and $h = g_1^{z_1} g_2^{z_2}$,

- chooses a function H uniformly at random from a suitable family of universal one-way hash functions,

- forwards to $\mathcal{A}$ the public key (g_1, g_2, c, d, h, H).

Note that $\mathcal{D}$ has slightly altered the key generation process. Next, $\mathcal{D}$ presents $\mathcal{A}$ with the standard IND-CCA challenge. For this, $\mathcal{D}$ has to simulate a decryption oracle for $\mathcal{A}$:

- [Decryption Oracle Sim$-\mathcal{O}_\mathcal{D}$]: on input a ciphertext $c = (c_1, c_2, c_3, c_4)$, the algorithm $\mathcal{D}$

  - computes $\hat{\alpha} = H(c_1, c_2, c_3)$ and tests if $c_1^{x_1 + y_1 \hat{\alpha}} c_2^{x_2 + y_2 \hat{\alpha}} = c_4$,
  - if the above check fails, $\mathcal{D}$ outputs $\perp$, otherwise $e/(g_3^{z_1} g_4^{z_2})$.

To form the challenge ciphertext, $\mathcal{D}$ can proceed as follows.

- [Challenge Construction]: after receiving two messages m_0 and m_1 from $\mathcal{A}$, the algorithm $\mathcal{D}$ selects $b \in \{0, 1\}$ uniformly at random and outputs (g_3, g_4, e, v), where

$$e = g_3^{z_1} g_4^{z_2} m_b,$$
$$\alpha = H(g_3, g_4, e) \text{ and}$$
$$v = g_3^{x_1 + y_1 \alpha} g_4^{x_2 + y_2 \alpha}.$$

Now, depending on the input to the distinguisher $\mathcal{D}$, two possible situations can occur.

Case 1. The distinguisher's input comes from $\mathbf{D_1}$. In this case, as the two last elements of the input 4-tuple (g_1, g_2, g_3, g_4) have been selected uniformly at random from G, we may actually assume the input has the form $(g_1, g_2, g_1^{r_1}, g_2^{r_2})$, for some r_1, r_2 chosen uniformly at random from $\mathbb{Z}_q$. As the situation $r_1 = r_2$ occurs with negligible probability only, we may assume $r_1 \neq r_2$. Let us denote by $w = \log_{g_1}(g_2)$, the discrete logarithm of g_2 with respect to the base g_1, and thus $g_4 = g_1^{wr_2}$.

Now, what happens if an invalid ciphertext $c' = (c_1', c_2', c_3', c_4')$, different from the challenge ciphertext $c = (c_1, c_2, c_3, c_4)$, is queried to the decryption oracle Sim$-\mathcal{O}_\mathcal{D}$? As c' is invalid, one of the following must be the case:

- $(c_1', c_2', c_3') = (c_1, c_2, c_3)$, however $c_4' \neq c_4$, thus

$$(c_1')^{x_1 + y_1 \hat{\alpha}} (c_2')^{x_2 + y_2 \hat{\alpha}} \neq c_4',$$

and therefore the check performed by $\text{Sim}-\mathcal{O}_\mathcal{D}$ fails;

- $(c_1', c_2', c_3') \neq (c_1, c_2, c_3)$. Due to the choice of H from a target collision resistant hash function family, except with negligible probability we have $H(c_1', c_2', c_3') \neq H(c_1, c_2, c_3)$; let us thus assume this is actually the case. Now, the decryption oracle will reject unless the equation $(c_1')^{x_1 + y_1 \alpha'} (c_2')^{x_2 + y_2 \alpha'} \neq c_4'$ holds, where α' stands for $H(c_1', c_2', c_3')$. Note that this equation will hold if and only if the point (x_1, x_2, y_1, y_2) belongs to the hyperplane $\overline{h}$ given by

$$\log_{g_1}(c_4') = r_1 x_1 + w r_2 x_2 + \alpha' r_1 y_1 + \alpha' r_1, \tag{5.1}$$

which can be argued to happen with negligible probability only: the point (x_1, x_2, y_1, y_2) is, in the eyes of $\mathcal{A}$, any point selected uniformly at random from the line $\mathcal{L}$ defined by the intersection of hyperplanes

$$\log_{g_1}(c) = x_1 + w x_2,$$
$$\log_{g_1}(d) = y_1 + w y_2 \text{ and}$$
$$\log_{g_1}(c_4') = r_1 x_1 + w r_2 x_2 + \alpha' r_1 y_1 + \alpha' w r_2 y_2.$$

However, this line only hits the hyperplane $\overline{h}$ at one point, and as a consequence Equation (5.1) above will only hold with negligible probability.

We have thus argued that all invalid ciphertexts will be rejected by $\text{Sim}-\mathcal{O}_\mathcal{D}$. Now, let us see that the distribution of the bit b is independent from $\mathcal{A}$'s view.

Consider the point $(z_1, z_2) \in \mathbb{Z}_q \times \mathbb{Z}_q$. From the public key, $\mathcal{A}$ can only infer that this is a uniformly at random chosen point from the line

$$\log_{g_1}(h) = z_1 + w z_2. \tag{5.2}$$

Note that h is the only value $\mathcal{A}$ knows in the above equation. All other information $\mathcal{A}$ may gather should come from answers provided by $\text{Sim}-\mathcal{O}_\mathcal{D}$, which, as being derived from valid ciphertext inputs, will only provide him with relations of the form $r' \log_{g_1}(h) = r' z_1 + r' w z_2$, which of course adds nothing to his knowledge. Moreover, the challenge ciphertext $c = (c_1, c_2, c_3, c_4)$ is constructed as $c_3 = c_1^{z_1} c_2^{z_2} m_b$, and the equation

$$\log_{g_1}(c_1^{z_1} c_2^{z_2}) = r_1 z_1 + w r_2 z_2.$$

is linearly independent from Equation (5.2). Thus no information about the "mask" $c_1^{z_1} c_2^{z_2}$ can possibly be obtained by $\mathcal{A}$, which completes our argument; we have thus proved that if the simulator input comes from $\mathbf{D_1}$ the distribution of the bit b is independent from $\mathcal{A}$'s view.

Case 2. The distinguisher's input comes from $\mathbf{D_2}$. Let us start by observing that the output distribution of challenges coming from $\mathcal{D}$ has exactly the same distribution as that induced by choosing b uniformly at random and encrypting m_b: indeed, the simulator defines $c_1 = g_3 = g_1^r$, $c_2 = g_4 = g_1^{wr}$, and computes $c = (g_1^r)^{x_1}(g_2^r)^{x_2}$, $d = (g_1^r)^{y_1}(g_2^r)^{y_2}$ and $h = (g_1^r)^{z_1}(g_2^r)^{z_2}$. As, moreover, $(g_1^r)^{x_1}(g_2^r)^{x_2} = c^r$, $(g_1^r)^{y_1}(g_2^r)^{y_2} = d^r$ and $(g_1^r)^{z_1}(g_2^r)^{z_2} = h^r$, one has $c_3 = m_b h^r$ and $c_4 = c^r d^{r\alpha}$, where $\alpha = H(c_1, c_2, c_3)$.

Now, the output distribution of $\text{Sim}-\mathcal{O}_{\mathcal{D}}$ is statistically indistinguishable from the output distribution of the IND-CCA decryption oracle $\mathcal{D}_{\mathcal{O}}$: from the above, it is easy to see that on input a valid ciphertext both oracles give exactly the same answer. Moreover, they reject all invalid ciphertexts except with negligible probability. Let us argue the latter, again by considering the distribution of a point $(x_1, x_2, y_1, y_2) \in \mathbb{Z}_q^4$ in the view of $\mathcal{A}$. Given the information $\mathcal{A}$ has, the point (x_1, x_2, y_1, y_2) has in his view been chosen uniformly at random from the intersection of the hyperplanes

$$\log_{g_1}(c) = x_1 + wx_2 \text{ and } \log_{g_1}(d) = y_1 + wy_2. \tag{5.3}$$

Now, suppose we have an invalid ciphertext $(c_1, c_2, c_3, c_4) \in G^4$, and that $c_1 = g_1^{r_1}$ and $c_2 = g_1^{wr_2}$, with $r_1 \neq r_2$. Then, both oracles $\text{Sim}-\mathcal{O}_{\mathcal{D}}$ and $\mathcal{O}_{\mathcal{D}}$ will reject unless

$$\log(v) = r_1 x_1 + wr_2 x_2 + H(c_1, c_2, c_3)r_1 y_1 + H(c_1, c_2, c_3)r_2 y_2,$$

which will only happen if actually (x_1, x_2, y_2, y_2) lies in the intersection line of the above hyperplane with the plane defined above by Equation (5.3). Thus, the first invalid query will be rejected with probability $1 - 1/q$. Furthermore, the ith invalid query will be rejected with probability at least

$$1 - \frac{1}{(q - i + 1)},$$

which concludes our argument.

Therefore, in this case the joint distribution of b and $\mathcal{A}$'s view are statistically indistinguishable from the joint distribution derived from a real attack scenario.

Thus, repeating the above simulation a number of times, the distinguisher can decide whether the input (g_1, g_2, g_3, g_4) was indeed a Diffie–Hellman tuple or not, just by counting the number of times $\mathcal{A}$ guessed the correct bit b. If he succeeded more than half of the times, the distinguisher will recognize the input as coming from $\mathbf{D_2}$, otherwise, it will decide it comes from $\mathbf{D_1}$.

$\square$

5.2 Going beyond: Tools

The core collision-resistance property to prove the security of Cramer and Shoup's public-key encryption scheme from CRYPTO '98 were target collision hash families from Definition 5.2. Now we are going to assume a different tool is at hand, namely a family of hash functions which can help us simulate a uniform distribution on a given set, and which, moreover, come with a *trapdoor* in some sense. These families of functions are, together with so-called *hard subset membership problems*, the tools underlying a general construction (again due to Cramer and Shoup) of a group-theoretic encryption scheme [CS02] which we will expound next in Section 5.3.

5.2.1 Projective hash families

We will need a few basic statistical notions measuring to what extent two probability distributions are similar,

DEFINITION 5.3 [Statistical distance and indistinguishability]
Let ξ and ν be random variables taking values in a finite set X. The statistical distance between ξ and ν is defined to be

$$\delta(\xi, \nu) = \frac{1}{2} \sum_{x \in X} |Pr[\xi = x] - Pr[\nu = x]|.$$

For a given $\varepsilon > 0$, we say that ξ and ν are ε-close if $\delta(\xi, \nu) \leq \varepsilon$. Furthermore, if $\xi = (\xi_\ell)_{\ell \geq 0}$ and $\nu = (\nu_\ell)_{\ell \geq 0}$ are sequences of random variables (where for each ℓ the corresponding ξ_ℓ and ν_ℓ takes values on the same finite set X_ℓ) we say that they are statistically indistinguishable if $\delta(\xi_\ell, \nu_\ell)$ is a negligible function in ℓ.

Similarly, if for any probabilistic polynomial time algorithm $\mathcal{A}$,

$$|\Pr[\mathcal{A}(\xi_\ell) = 1] - \Pr[\mathcal{A}(\nu_\ell) = 1]|$$

is negligible in ℓ, we refer to the sequences as computationally indistinguishable.

Now we can formally define the basic cryptographic tools we will need in this section and look at some simple examples.

DEFINITION 5.4 [Projective hash function]
By X, Π, and S we denote finite, non-empty sets, and consider a distinguished subset $L \subseteq X$. Let K be a finite index set, and $\mathbf{H} = \{H_k\}_{k \in K}$ a collection of mappings with domain X and range in Π. Moreover, let $\alpha : K \longrightarrow S$ be a

mapping. We say that $\mathcal{H} = (\mathbf{H}, K, X, L, \Pi, S, \alpha)$ *is a* projective hash family
for (X, L)—*in short, a* PHF—*if for all* $k_1, k_2 \in K$ *and for all* $x \in L$ *the
equality* $H_{k_1}(x) = H_{k_2}(x)$ *holds if and only if* $\alpha(k_1) = \alpha(k_2)$.

The above definition can be stated equivalently by saying that for all $k \in K$
the action of H_k on L is completely determined by $\alpha(k)$. Note, however, that
this does not imply that $\alpha(k)$ suffices to compute $H_k(x)$ for any $x \in L$;
additional information may be required in order to do so.

Example 5.1

Let G be a finite non-Abelian group, fix $x \in G$ and consider $L = C_G(x)$, i.e.,
L is comprised of exactly those group elements $y \in G$ which satisfy $xy = yx$.
Now, consider a finite index set $K \subseteq \mathbb{N}$ and define for each $k \in K$ a map
$H_k : G \longrightarrow G$ by $H_k(a) = a^{-k}x^k a^{k+1}$. Moreover, take $\alpha : K \longrightarrow G$ defined
by $\alpha(k) = x^k$. Clearly, we have for all $g \in L$ that $H_k(g) = \alpha(k)g$, and thus
the above $\mathcal{H} = (\{H_k\}_{k \in K}, K, G, L, G, G, \alpha)$ is a projective hash family. □

Example 5.2

Let V be a finite-dimensional vector space over a finite field K. Given a basis
$\{v_1, \ldots, v_{\dim(V)}\}$ for V, let L be the proper subspace spanned by $v_1, \ldots, v_t$ for
some $t < \dim(V)$. Now, for each $k \in K$ define $H_k : V \longrightarrow V$ as $H_k(v) = kv$,
and furthermore, define $\alpha : K \longrightarrow L^t$ as $(kv_1, \ldots kv_t)$. It is easy to see that
$\mathcal{H} = (\{H_k\}_{k \in K}, K, V, L, V, L^t, \alpha)$ is a projective hash family. □

As can be seen from the above examples, there are cases in which the
projection α discloses a lot of information (sometimes even enough) on how
to compute $\mathbf{H}$-images of elements not lying in L. However, as we will see,
the cases which are cryptographically more relevant involve projective hash
families such that the projection does not help in evaluating $\mathbf{H}$ for elements
$x \in X$, unless one has computed some *evidence* that $x \in L$. Depending on
the amount of information the projection discloses about the behavior of $\mathbf{H}$
outside L, projective hash families are called:

- ε-*universal* if for any $x \in X \setminus L$ and for $k \in K$ (chosen uniformly at
  random), the probability of correctly guessing $H_k(x)$ from x and $\alpha(k)$
  is at most ε.

- ε-universal$_2$ if for any $x, x^* \in X \setminus L$, $x \neq x^*$, for $k \in K$ chosen uniformly
  at random, the probability of correctly guessing $H_k(x)$ from $\alpha(k)$ and
  $H_k(x^*)$ is at most ε.

- ε-*smooth* if the probability distributions of $(x, s, H_k(x))$ and (x, s, π),
  where k, x, and π are chosen uniformly at random in K, $X \setminus L$, and Π,

respectively, and $s = \alpha(k)$, are ε-close.[1]

- *strongly universal$_2$* if for $k \in K$ chosen uniformly at random, for any $x, x^* \in X \setminus L$, $x \neq x^*$ the random variables

  – $\xi_k = H_k(x)$, conditioned to $\alpha(k)$,
  – η_k, the variable ξ_k conditioned to both $\alpha(k)$ and $H_k(x^*)$,

  are statistically close to the uniform distribution over Π.

For the sake of readability, we will often drop the prefix "ε" in the sequel, when this yields no ambiguity.

The above notions provide a way of quantifying how much information about the behavior of **H** outside L is leaked by the projection map: if the family is universal, $\alpha(k)$ hardly gives any information about $H_k|_{X \setminus L}$. If **H** is universal$_2$, not even knowing, besides the projection given by α, some information about the behavior of H on a fixed (arbitrary) point of $X \setminus L$ will help on guessing its action on a new point $x \in X \setminus L$. Smoothness is achieved if, on average (taken over $k \in K$, $x \in X \setminus L$), guessing $H_k(x)$ given α is no better than a random guess. Strongly universal$_2$ is roughly speaking worst-case smoothness, i.e., fixing any $x \in X \setminus L$, the value $H_k(x)$ is (close to) a uniform random variable on Π, whose distribution is induced by choosing k uniformly at random. Moreover, this still holds if the behavior of H on a given point of $X \setminus (L \cup \{x\})$ is known. It is easy to see that strongly universal$_2$ PHFs are smooth, and similarly universal$_2$ PHFs are also universal. That is, roughly speaking, strongly universal$_2$ PHFs are the ones in which the projection gives least information about how $\{H_k\}_{k \in K}$ acts on $X \setminus L$, whereas given a PHF that can only be proven universal it may be the case that α leaks relevant information of what goes on outside L. However, there are efficient generic methods to "upgrade" the weaker types of PHFs to achieve more robust constructions (see [CS02]).

REMARK 5.2   In this section, the described properties of projective hash families have been introduced in an information-theoretic scenario—in the sense that they do not depend on the (hypothetical) hardness of any computational problem. Sometimes we will make use of computational versions of them, that is, the desired hardness properties will only hold under certain restrictions.   ☐

---

[1] A stronger notion of smoothness (which we will refer to as *worst-case smoothness*) may be defined by imposing that, for any $x \in X \setminus L$ the distribution of $H_k(x)$ conditioned to s is ε-close to the uniform distribution over Π. This is actually achieved in most cases.

5.2.2 Subset membership problems

Another tool we need in Cramer and Shoup's general construction are so-called *subset membership problems*, which naturally capture indistinguishability properties—the latter lying at the heart of a number of cryptographic contructions. To avoid getting very technical in this introductory text, in the sequel we only give a somewhat informal description of what is meant by a subset membership problem, and also describe informally how they can link together with projective hash families in what are called *hash proof systems*.

Many cryptosystems base their security in the sense of IND-CPA or IND-CCA on a decisional assumption, the decisional Diffie–Hellman assumption which we saw earlier (Definition 4.1) being a prominent example. Another example is the quadratic residuosity (QR) assumption which captures the task of deciding whether an element of the unit group $\mathbb{Z}_N^*$ with a composite natural number N has a root. Many of these problems can be formulated in terms of indistinguishability of two probability distributions.

Since computational assumptions are complexity theoretic statements in nature, a complexity parameter $\ell \in \mathbb{N}$, is taken into account. Also, for each value of ℓ, we can have several possible instances of the same problem; the (random) choice of a particular instance for complexity parameter ℓ is modeled by a probability distribution I_ℓ. To describe a subset membership problem, we also consider a probabilistic polynomial time algorithm, called the *instance generator*, which on input 1^ℓ, outputs a description i of a set X_i and a subset $L_i \subset X_i$. The instance generator also outputs a *witness set* W_i whose elements provide "proofs of belonging" to the elements in L_i, that is, given $x \in L_i$, there is always a $w \in W_i$ that can be used to prove that x belongs to L_i. Thus, there is a binary relation $\mathcal{R} \subseteq L_i \times W_i$ inherent to each instance (described by) i.

Given the above, we are ready to specify a *subset membership problem* $\mathcal{M}$ by means of the collection of distributions $(I_\ell)_{\ell \in \mathbb{N}}$ together with some sampling and verification algorithms (see [CS02]):

- sampling according to the distribution I_ℓ,

- deciding, on input an instance i, whether a given input element x is actually in X_i,

- selecting uniformly at random, on input an instance i, an element $x \in L_i$ and a witness $w \in W_i$ for it.

A subset membership problem $\mathcal{M}$ is *hard* if the probability distributions (i, x) and (i, x'), where i is the output of the instance generator and x, x' are uniformly distributed on L_i and $X_i \setminus L_i$, respectively, are polynomially indistinguishable. The following example relates the encryption scheme discussed in the previous section to the hardness of a subset membership problem.

Example 5.3
For each $\ell \in \mathbb{N}$ consider the distribution I_ℓ induced by the following selection process: take uniformly at random an ℓ-bit prime p and a generator g_1 of a cyclic group G of order p. Moreover, also select uniformly at random an integer $a \in \{0, \ldots, p-1\}$. We set $g_2 = g_1$. The complete description of this instance is then $X = G \times G$, $L = \langle (g_1, g_2) \rangle$ (i. e., the "diagonal subgroup" of $G \times G$ generated by (g_1, g_2) and $W = \{0, \ldots, p-1\}$. Indeed, if an element $(x_1, x_2) \in G$ is also in L, then there must exist an exponent $w \in W$ so that $x_i = g_i^w$, for $i = 1, 2$. Thus, we may see I_ℓ as the distribution induced on the triplets

$$i = (g_1, g_2, \{0, \ldots, p-1\})$$

selected as above.

To complete the subset membership problem specification, one had to describe efficient algorithms for sampling according to the distribution I_ℓ, deciding membership in a corresponding X and selecting pairs $(x, w) \in L \times W$. For the sake of this example, let us assume that such algorithms are available. Then, the above subset membership problem is hard if and only if for each instance $i = (g_1, g_2, \{0, \ldots, p-1\})$, the corresponding distributions

$$(g_1, g_2, \{0, \ldots, p-1\}, x_1, x_2) \text{ and } (g_1, g_2, \{0, \ldots, p-1\}, x_1', x_2')$$

are polynomially indistinguishable, where (x_1, x_2) is chosen uniformly at random from L and (x_1', x_2') from $X \setminus L$. Note that the latter condition is exactly Assumption 5.1.   □

5.2.3   Hash proof systems

A *hash proof system* (HPS) provides a link between a subset membership problem and a collection of projective hash families. More precisely, each instance i of a hash proof system $\mathcal{P}$ will be specified by a corresponding instance (X_i, L_i, W_i) of a subset membership problem and a "matching" instance $(H_i, K_i, X_i, L_i, \Pi_i, S_i, \alpha_i)$ of a projective hash family—plus some efficient algorithms. For example, some of the algorithms provided by an instance i of $\mathcal{P}$ are:

- the *private evaluation algorithm*, that on inputs i, $k \in K$ and $x \in X$ outputs $H_k(x)$,

- a *sampling algorithm* for $L \times W$, that on input i outputs a random $x \in L$ and a witness $w \in W$ for x,

- the *public evaluation algorithm*, that on inputs i, $s \in S$, and $(x, w) \in \mathcal{R}$ outputs $H_k(x)$, for any $k \in K$ such that $s = \alpha(k)$.

Notions of (ε-)universality and (ε-)smoothness for $\mathcal{P}$ are directly inherited from those of the underlying projective hash families. However, now ε has to be seen as a function $\varepsilon = \varepsilon(\ell)$ of the complexity parameter. If this function is negligible, then the prefix *strongly* is added to the qualifiers "universal," "universal$_2$," or "smooth," respectively. The key observation in the general construction by Cramer and Shoup is the following.

REMARK 5.3   If a hash proof system is strongly universal and the underlying subset membership problem is hard, then the problem of evaluating $H_k(x)$ for random $k \in K$ and arbitrary $x \in X$ given only x and $\alpha(k)$ is also hard. Thus, the role of the witness in the public evaluation algorithm becomes clear: without w there is no way to efficiently compute $H_k(x)$.   ⬚

Working out the details of projective hash families, subset membership problems, and hash proof systems is somewhat technical. They are versatile building blocks, however, for the construction of encryption schemes as well as other cryptographic protocols—including hybrid encryption [KD04], password-authenticated key exchange [GL06] and oblivious transfer [TK05].

5.3   General Cramer–Shoup encryption scheme

Given a strongly smooth hash proof system for a hard membership problem, involving a projective hash family, from an instance $(H, K, X, L, \Pi, S, \alpha)$, the main idea to derive a public-key encryption scheme is the following: the secret key, which enables the decryption of ciphertexts, is $k \in K$ and the public key consists of the projection $s = \alpha(k)$ along with the instance description. The message space is the set Π, and to encrypt a message $m \in \Pi$, a random pair $(x, w) \in L \times W$ is generated, so that w is a witness of x. Next, the public evaluation algorithm is used to obtain $H_k(x)$. The ciphertext is the pair $(x, H_k(x) \cdot m)$, where $\cdot$ is a suitable binary operation; we assume that Π is contained in a group where elements can be efficiently inverted and multiplied, and that $\cdot$ is a group operation (such as XOR).

Clearly, the holder of k can retrieve $H_k(x)$ and thus the message by using the private evaluation algorithm. On the other hand, since the subset membership problem is hard, there is no way for a polynomially bounded adversary to distinguish between a well-formed ciphertext and a fake ciphertext obtained by choosing $x \in X \setminus L$ instead of $x \in L$. But, due to the smoothness of the hash proof system, since k is unknown, $H_k(x)$ is close to be uniformly distributed on Π, so the message is nearly perfectly hidden. Therefore, no information about the secret can be obtained in polynomial time by a passive adversary.

IND-CCA security is achieved by appending to the ciphertext a "proof of integrity" obtained from a strong universal$_2$ (extended) HPS. The set "extending" X and L in the definition of the latter HPS is just the message space Π. Also, new values are appended to the secret and public keys. More precisely, let $\mathcal{M}$ be a hard subset membership problem and $\mathcal{P}, \hat{\mathcal{P}}$ be two hash proof systems for $\mathcal{M}$, strongly smooth and strongly universal$_2$ extended, respectively. Then the Cramer–Shoup encryption scheme is described by Figure 5.2.

$\mathcal{K}$: on input 1^ℓ, this algorithm

  - selects appropriate instances of $\mathcal{P}, \hat{\mathcal{P}}$ and $\mathcal{M}$.
    The instances of $\mathcal{P}$ and $\hat{\mathcal{P}}$ must share the sets X, L, and W and the sampling algorithm. These consist of a triplet (X, L, W) associated to $\mathcal{M}$ and two tuples $\mathbf{H} = (H, K, X, L, \Pi, S, \alpha)$ and $\hat{\mathbf{H}} = (\hat{H}, \hat{K}, X \times \Pi, L \times \Pi, \hat{\Pi}, \hat{S}, \hat{\alpha})$ of $\mathcal{P}$ and $\hat{\mathcal{P}}$, respectively.
  - chooses $k \in K$ and $\hat{k} \in \hat{K}$ uniformly at random.
  - computes $s = \alpha(k) \in S$, $\hat{s} = \hat{\alpha}(\hat{k}) \in \hat{S}$.
  - outputs the public key, which is a description of the involved instances plus the tuple $(s, \hat{s})$, and the private key $(k, \hat{k})$.

$\mathcal{E}$: on input (pk, m), where $m \in \Pi$, generates $x \in L$ and a corresponding witness $w \in W$ by means of the sampling algorithm.

Then, $\mathcal{E}$

  - computes $\pi = H_k(x)$, (from x, s, and w by using the public evaluation algorithm provided by $\mathcal{P}$),
  - $e = m \cdot \pi \in \Pi$ and $\hat{\pi} = \hat{H}_{\hat{k}}(x, e)$ (from $\hat{s}$, x, e, and w by using the public evaluation algorithm provided by $\hat{\mathcal{P}}$).

The ciphertext is the tuple $(x, e, \hat{\pi})$.

$\mathcal{D}$: on input the secret key $(k, \hat{k})$ and a ciphertext $(x, e, \hat{\pi})$, this algoritm

  - computes $\hat{\pi}' = \hat{H}_{\hat{k}}(x, e) \in \hat{\Pi}$ (by means of the private evaluation algorithm of $\hat{\mathcal{P}}$),
  - checks if $\hat{\pi} = \hat{\pi}'$ and, if not, aborts with the output $\perp$. Otherwise, $\mathcal{D}$ computes $\pi = H_k(x) \in \Pi$ (using the private evaluation algorithm of $\mathcal{P}$) and outputs the plaintext $m = e \cdot \pi^{-1} \in \Pi$.

This algorithm is also supposed to recognize and reject bitstrings that do not encode an element of $X \times \Pi \times \Pi$.

FIGURE 5.2:   Cramer–Shoup—general construction.

5.4 A concrete instantiation

We have already seen a scheme derived from the general construction above—the Cramer–Shoup encryption from 1998 which was the topic of Section 5.1. This derivation has actually some subtle points the interested reader may want to follow up with [CS02]. Below we outline another instantiation from a hardness assumption which we have not discussed so far, the so-called *decision composite residuosity assumption*.

Assumption 5.2 (Decision composite residuosity assumption) *Let $p' \neq q'$ be odd Sophie Germain primes with $p = 2p' + 1$ and $q = 2q' + 1$ having the same bitlength. Set $N = p \cdot q$ and let P be the subgroup of all N^{th} powers of elements in $\mathbb{Z}_{N^2}^*$. Then the uniform distributions on P and $\mathbb{Z}_{N^2}^*$ are computationally indistinguishable.*

The above assumption was first introduced by Pascal Paillier who proposed a public-key cryptosystem based on it [Pai99], though he stated it only requiring p and q to be prime. Let us see how Cramer and Shoup defined a subset membership problem from this assumption, not using $\mathbb{Z}_{N^2}^*$ and P themselves but two convenient (cyclic) subgroups:

On input 1^ℓ, our instance generator selects two ℓ-bit primes p and q as above. Then the group $\mathbb{Z}_{N^2}^*$ can be written as a direct product

$$\mathbb{Z}_{N^2}^* = C_N C_{N'} C_2 T, \tag{5.4}$$

where $N' = p'q'$, each C_m is a cyclic group of order m and T is the subgroup of $\mathbb{Z}_{N^2}^*$ generated by $-1 \mod N^2$. Now, it is not hard to see that the subgroup $X = C_N C_{N'} T$ is cyclic of order $2NN'$, and $L = C_{N'} T$, the subgroup of Nth powers of X, is cyclic of order $2N'$. The output of the instance generator consists of N, a random generator g for L and the witness set $W = \{1, \ldots, \lfloor \frac{N}{2} \rfloor\}$; obviously, $w \in W$ will be a witness for $x \in X$ if $x = g^w$.

Assumption 5.2 implies that it is hard to distinguish $X \setminus L$ from L: if x is selected uniformly at random from $\mathbb{Z}_{N^2}^*$ (respectively P) and so is b from $\{0, 1\}$, then $x \cdot (-1)^b$ is uniformly distributed in X (respectively L). Hence, distinguishing X from L is no easier than distinguishing $\mathbb{Z}_{N^2}^*$ from P, i.e., Assumption 5.2 implies it is hard to distinguish X from L, and so it is to distinguish $X \setminus L$ from L (note that $|X \setminus L|/|X|$ goes to 1 as ℓ goes to infinity).

Next, we need two hash proof systems for our subset membership problem, one strongly smooth and another one strongly universal$_2$ extended. The construction of these is technically rather involved and here we will only argue informally how they can be derived for a concrete instance description specifying X, L, and g as above. Let $\hat{\mathbf{H}}_* = (\hat{H}, \hat{K}_*, X, L, X, L, \alpha)$ be constructed as follows:

- $\hat{H} = \text{Aut}(X)$;

- $K_* = \{0, \ldots, 2NN' - 1\}$—as X is cyclic of order $2NN'$, each of these integers defines an automorphism of X, and vice versa, so each automorphism of X can be indexed by a value k (thus, we take $H_k(x) = x^k$);

- α maps each $k \in \mathbb{Z}$ to $H_k(g) = g^k \in L$.

It can be proved that the above family is $2^{-\ell}$−universal (for ℓ the bit size of p and q). As N' is not public, we have to make a slight modification and take the set $K = \{0, \ldots, \lfloor \frac{N^2}{2} \rfloor\}$ instead of K_*. This modification is not relevant for our purposes, and the projective hash family $\hat{\mathbf{H}} = (\hat{H}, K, X, L, X, L, \alpha)$ will suffice to construct a strongly universal$_2$ extended hash proof system as required (using the extension methods specified in [CS02]).

Finally, we also need a strongly smooth hash proof system. For this, consider the map $\chi : \mathbb{Z}_{N^2} \longrightarrow \mathbb{Z}_N$ defined by $\chi(a + bN \mod N^2) = b \mod N$, where $0 \le a, b, < N$. It is easy to see that its restriction to any coset of C_N is a one-to-one map from that coset onto $\mathbb{Z}_N$. Now, take $H = \chi \circ \hat{H}$. Then it is possible to argue that the corresponding projective hash family $\mathbf{H} = (H, K, X, L, X, L, \alpha)$ is strongly smooth, and the resulting encryption scheme is depicted in Figure 5.3.

5.5 Projective hash families from (non-Abelian) groups

In search of projective hash families that are suitable for their general construction, Cramer and Shoup identified a tool based on Abelian groups—so-called *group systems* [CS02]. They were able to give generic construction methods for deriving projective hash families that could be proved to fulfill the desired (universality and smoothness) properties, provided that the underlying group systems were "good enough." With the aim of allowing non-Abelian groups into play, we here give a brief description of a somewhat generalized version of Cramer and Shoup's approach, as it was put forward in [GVV08]. A special case of this generalization, as well as a corresponding non-Abelian version of the general Cramer and Shoup encryption scheme, can be found in [GVMSV05].

5.5.1 Group action systems

Let us start by defining *group action systems*, which generalize the notion of *group system* as introduced by Cramer and Shoup. For this, denote by X a finite set and consider a finite (not necessarily Abelian) group H left-acting on X. Thus, each element $\phi \in H$ can be seen as an element of the symmetric group on X, and for all $\phi_1, \phi_2 \in H$ and $x \in X$ we have $(\phi_1 \phi_2)x = \phi_1(\phi_2 x)$.

$\mathcal{K}$: on input 1^ℓ, this algorithm

- selects appropriate instances of the corresponding HPSs, both for the same value of $N = p \cdot q$. The message space is $\mathbb{Z}_N$, and an encoding map $\Gamma : \mathbb{Z}_{N^2} \times \mathbb{Z}_N \longrightarrow \{0, \ldots, 2^\ell - 1\}^n$ for a suitable n is published;

- chooses a random generator g for L;

- chooses $k, \widetilde{k}, \hat{k}_1, \ldots, \hat{k}_n \leftarrow K$ and computes $s = g^k, \widetilde{s} = g^{\widetilde{k}}, \hat{s}_i = g^{k_i}$, for $i = 1, \ldots, s$;

- outputs the public key $(g, s, \widetilde{s}, \hat{s}_1, \ldots, \hat{s}_n)$, all public parameters and the private key $(k, \widetilde{k}, \hat{k}_1, \ldots, \hat{k}_n)$.

$\mathcal{E}$: to encrypt $m \in \mathbb{Z}_N$ given the public key and parameters,

- choose $w \leftarrow W$, compute $x = g^w, y = s^w, \pi = \chi(y), e = m + \pi$,

- set $\Gamma(x, e) = (\gamma_1, \ldots, \gamma_n)$,

- compute $\hat{y} = \widetilde{s}^w \prod_{i=1}^n \hat{s}_i^{\gamma_i w}$, $\hat{\pi} = \chi(\hat{y})$, and output $(x, e, \hat{\pi})$.

$\mathcal{D}$: on input the secret key and a ciphertext $(x, e, \hat{\pi})$,

- set $\Gamma(x, e) = (\gamma_1, \ldots, \gamma_n)$, and compute $\hat{y} = x^{\widetilde{k} + \sum_{i=1}^n \gamma_i \hat{k}_i}$ and $\hat{\pi}' = \chi(\hat{y})$,

- check if $\hat{\pi} = \hat{\pi}'$ and, if not, abort with output $\bot$. Otherwise, compute $y = x^k, \pi = \chi(x) \in \Pi$ and output $m = e - \pi$.

FIGURE 5.3: Cramer and Shoup encryption from Assumption 5.2.

Next, let S be some finite group and $\chi : H \longrightarrow S$ a group homomorphism. Then, for any $\phi \in H$, the image $\chi(\phi)$ gives some (limited) information about ϕ, and thus χ provides partial information about the action of H on X. This partial information will eventually play the role of the information given by the mapping α in the projective hash families we will be constructing. Consequently, in our construction the action of group elements $\phi \in \ker \chi$ will be those for which the projection α will be less informative. Furthermore, if for a given point $x \in X$, we have that all elements from $\ker \chi$ fix it, then the value $\chi(\phi)$ determines the value of $\phi(x)$ for all $\phi \in H$, i.e., in this case the action of H on x is fully determined by χ.

DEFINITION 5.5 [Group action system] *Let X, H, S, and χ be defined as above. Then the tuple (X, H, χ, S) is called a* group action system.

As just noted, what remains "obscure" having only χ at hand is the action of $\ker \chi$ on X, which is defined by the action of H. For every $x \in X$, we denote by $[x]$ the $(\ker \chi)$-orbit of x, i.e.,

$$[x] = \{\phi x \mid \phi \in \ker \chi\}.$$

The action of H on the set of points that remain fixed by $\ker \chi$ is completely determined by χ. On the other hand, those $x \in X$ with large $(\ker \chi)$-orbits are the elements for which, knowing χ, less is known about how H acts on them. We define a subset L of X, which is comprised of all those elements in X whose orbit under $\ker \chi$ consists of a single element:

$$L = \{x \in X \mid |[x]| = 1\}$$

Equivalently, we can write $L = \{x \in X \mid [x] = \{x\}\}$.

Observe that the elements in $\ker \chi$ stabilize L, i.e., $\ker \chi \subseteq \text{Stab}(L)$, although these two sets are not necessarily equal. Also, H leaves L invariant, as for any $\phi \in H$ and $x \in L$, the element ϕx is fixed by all $\psi \in \ker \chi$. The latter holds, because there exists $\rho \in \ker \chi$ such that $\psi \phi = \phi \rho$ and thus $\psi \phi x = \phi \rho x = \phi x$.

As our aim is to construct cryptographically useful projective hash families, the systems above will be useful for us if χ gives little information about the action of H on $X \setminus L$. Thus, we will be particularly interested in those systems for which the $(\ker \chi)$-orbits of elements in $X \setminus L$ are large. To capture this idea, the following notion of diversity is helpful.

DEFINITION 5.6 [p-diversity]   *Let $p > 1$ be a positive integer. Then the group action system (X, H, χ, S) is called p-diverse if $|[x]| \geq p$ for all $x \in X \setminus L$.*

A trivial observation already tells us whether a group system is p-diverse for some values p, in particular for p being the smallest prime dividing $|H|$.

LEMMA 5.1
Let (X, H, χ, S) be a group action system, and let p be the smallest prime dividing $|\ker \chi|$. Then (X, H, χ, S) is p-diverse.

PROOF   Note that $\ker \chi$ acts on X, and thus $|[x]|$ divides $|\ker \chi|$. So if $x \in X \setminus L$ (i.e., if $|[x]| \neq 1$) then $|[x]|$ is at least p.   ☐

To make our discussion more concrete, let us take a look at some simple examples of group systems:

Example 5.4 [Cramer and Shoup's Abelian group systems]
As mentioned in the beginning, a *group action system* as we defined it, is inspired by—and generalizes—the notion of *group system*, as discussed by

Cramer and Shoup: given three finite Abelian groups X, L and Π, where L is a proper subgroup of X, a *group system* is just a tuple $(\mathcal{H}, X, L, \Pi)$, where $\mathcal{H} \leq \mathrm{Hom}(X, \Pi)$ is any chosen subgroup of the homomorphism group between X and Π.

Now, for any $h \in \mathrm{Hom}(X, \Pi)$ the mapping $\phi(x, \pi) = (x, h(x)\pi)$ is actually an element of $\mathrm{Aut}(X \times \Pi)$. Further, denote by Ψ the group monomorphism

$$\begin{array}{rcl} \Psi : \mathrm{Hom}(X, \Pi) & \longrightarrow & \mathrm{Aut}(X \times \Pi) \\ h & \longmapsto & \phi \end{array},$$

and consider the natural restriction

$$\begin{array}{rcl} \eta : \mathrm{Hom}(X, \Pi) & \longrightarrow & \mathrm{Hom}(L \times \Pi) \\ h & \longmapsto & h_{|L} \end{array}.$$

Finally, let $\chi = \eta \circ \Psi^{-1}$ be the functional composition of these two homomorphisms. Then, putting it all together, we obtain a group action system $(X \times \Pi, \Psi(H), \chi, \mathrm{Hom}(L \times \Pi))$.

□

Example 5.5 [Making use of collineations]
Let π be a finite projective plane over a finite prime field $\mathbb{F}_q$. Automorphisms of π are usually called *collineations*, and it is easy to see that the action of a collineation is faithful both on the point-set and on the line-set of π. (For basic notions on collineation groups of finite planes, see [Bon99].)

An *elation* is a collineation fixing both each point of a fixed line l (called the *axis*) and each line through a point C in l; the point C is usually referred to as the *center*. The fixed points of a non-identical elation are exactly the points on the axis, while the only fixed lines are those incident in the center. Let X be the point-set of π, L a fixed line in π, and C a fixed point on the line L. Take H as the group consisting of all elations with center C.

Notice that every elation in H induces a permutation of the points of L, since the center C is contained in L. We define χ as the group homomorphism translating each elation in H into the corresponding permutation of L:

$$\begin{array}{rcl} \chi : H & \longrightarrow & S_L \\ \zeta & \longmapsto & \zeta_{|L} \end{array}$$

Clearly, $\ker \chi$ is exactly the subgroup of all elations in H with axis L. Note that, moreover, the points in L are exactly the points of X stabilized by $\ker \chi$. On the other hand, if A is an arbitrary point outside L, each elation in $\ker \chi$ is uniquely determined by the image of A, that can be any point in the line defined by C and A, except for C (as an elation is completely determined giving its center, axis, and the image of a point outside the axis). Hence, $|\ker \chi| = q$ and therefore q-diversity is guaranteed in this construction. □

5.5.2   Group action projective hash families

Next, we outline the construction of a projective hash family from a group action system, following [CS02, GVMSV05]. The idea of this construction is to exploit a nice property (p-diversity) of the group action system to obtain nice properties of the projective hash family (universality and smoothness).

Consider a group action system (X, H, χ, S), and denote by $\hbar : K \to H$ a bijection from a suitable index set K (which will later serve as the private key space). Noting that $\chi(\hbar(k))$ determines the action of $\hbar(k)$ on the set L completely, it is easy to see that the tuple $(H, K, X, L, X, S, \chi \circ \hbar)$ is a projective hash family.

DEFINITION 5.7 [Group action projective hash family]   *Any projective hash family constructed from a group action system as described above is called a* group action projective hash family*—in short, an AcPHF. Such a projective hash family is made explicit by the tuple* $(X, H, K, S, \chi, \hbar)$.

To how diversity of the group action system translates into nice properties of the corresponding AcPHF, let us look at universality. First, we verify that for any $x \in X$, choosing $k \in K$ uniformly at random (that is, choosing uniformly at random a group element in H), given $\chi(\hbar(k))$, there are exactly $||[x]||$ equally probable candidates for $\hbar(k)x$.

LEMMA 5.2
Let (X, H, χ, S) be a group action system and let $x \in X$. If $\phi \in H$ is chosen uniformly at random, once $s = \chi(\phi)$ is given then ϕ is uniformly distributed on the coset $\chi^{-1}(s)$ and ϕx is uniformly distributed on the set $\{\psi x \mid \psi \in \chi^{-1}(s)\}$, that is, on a set of cardinality equal to $||[x]||$.

PROOF   Because of ϕ being chosen uniformly at random, once we fix $s = \chi(\phi)$, the resulting distribution is uniform on $\chi^{-1}(s)$. Moreover, for any $x \in X$, ϕx is uniformly distributed on

$$\{\psi x \mid \psi \in \chi^{-1}(s)\}$$

provided that the sets

$$S_y = \{\psi \in \chi^{-1}(s) \mid \psi x = y\}$$

are of the same size for all $y \in \{\psi x, \psi \in \chi^{-1}(s)\}$. But this is straightforward to see, as all S_y are left-cosets modulo $\ker \chi \cap \mathrm{Stab}(\{x\})$.   □

As a result, we know that for a given x, the size of its $\ker \chi$-orbit $[x]$ is our measure for the information χ provides on the action of H on x. In other words, we have the following result:

PROPOSITION 5.1

Let $\mathbf{H} = (X, H, K, S, \chi, \hbar)$ be a group action projective hash family. If the underlying group action system (X, H, χ, S) is p-diverse then the group action system $\mathbf{H}$ is $1/p$-universal.

PROOF   From Lemma 5.2, for any $x \in X \setminus L$, the probability of guessing the correct value of $\hbar(k)x$ for a random choice of $k \in K$ given $\chi(\hbar(k))$ is $1/|[x]|$, that is at most $1/p$. $\Box$

Once a universal projective hash family is constructed choosing the right elements according to the results above, the generic method of [CS02] may be used in order to transform it into a smooth projective hash family. Moreover, this transformation is sometimes superfluous: in some special cases, smoothness can even be guaranteed directly.

PROPOSITION 5.2

Let $\mathbf{H} = (X, H, K, S, \chi, \hbar)$ be a group action projective hash family. If the whole set $X \setminus L$ is a single orbit under the action of $\ker \chi$ then the group action projective hash family $\mathbf{H}$ is $|L|/|X|$-smooth.

PROOF   Let $x \in X \setminus L$. From Lemma 5.2, $\hbar(k)x$ is uniformly distributed on a set of size $|[x]| = |X \setminus L|$. Then, the statistical distance between $\hbar(k)x$ and the uniform distribution on X is

$$\frac{1}{2} \sum_{x \in X \setminus L} \left| \frac{1}{|X| - |L|} - \frac{1}{|X|} \right| + \frac{1}{2} \sum_{x \in L} \frac{1}{|X|} = \frac{|L|}{|X|}.$$

Therefore the probability distribution of $\hbar(k)x$ is $|L|/|X|$-close to the uniform distribution on X. $\Box$

Similarly, as from a universal PHF one can derive a smooth PHF, a simple extension of a universal PHF leads to the construction of a universal$_2$ PHF. Furthermore, in [CS02], Cramer and Shoup outline a generic transformation from any ε-universal projective hash family to an ε-universal$_2$ extended projective hash family. A more efficient extension can also be found in [GVV08]; both are simple yet rather technical.

Summarizing the discussion in this section, we can say that needed cryptographic hash proof systems may be derived from suitable group action systems: Cramer and Shoup's construction for an IND-CCA encryption scheme required a hard subset membership problem $\mathcal{M}$ and two suitable hash proof systems. Diverse group action systems also provide strongly universal$_2$ projective hash families for the Kurosawa–Desmedt encryption scheme [KD04].

As a result, it is possible to translate nice group-theoretic properties of simple building blocks into cryptographic robustness for practical purposes.

5.6   Summary and further reading

In this chapter we have

- presented the public-key encryption scheme of Cramer and Shoup from 1998, which offers IND-CCA security in the standard model, and detailed its corresponding security proof.

- introduced the basic tools behind the general Cramer and Shoup construction for achieving schemes with security in the sense of IND-CCA, namely *projective hash families*, *subset membership problems*, and the tool to combine them suitably: *hash proof systems*.

- sketched the general Cramer and Shoup construction from 2002, as well as a concrete instantiation of it.

- elaborated on ways to adapt the Cramer and Shoup framework for using non-Abelian groups as a base. As a main tool to this aim we introduced *group action systems*.

The main results and schemes from this chapter are taken from the original papers of Cramer and Shoup [CS98, CS02]. The non-Abelian adaptation comes from [GVMSV05, GVV08]. For the reader who is interested in reading about other applications of projective hash families within cryptography, the literature offers quite a number of illustrations of this technique; some nice examples are due to work by Kurosawa and Desmedt (hybrid encryption [KD04]), Gennaro and Lindell (password-authenticated key exchange [GL06]), Halevi and Kalai (oblivious transfer [TK05, HK12]), and Abdalla et al. (commitment schemes [ACP09]). This list already indicates that, despite the technical details that have to be taken care of, projective hash families have established themselves as an important tool in cryptographic protocol design.

Furthermore, in this chapter we have taken a look at the importance of different types of hash function families, which are nowadays at the core of many cryptographic constructions. One of the seminal papers pointing out the relevance of universal hashing is due to Naor and Yung [NY89].

5.7 Exercises

Exercise 34 *Change the encryption algorithm in Cramer–Shoup '98 by modifying the definition of α; namely, let*

$$\alpha = H(u_1, u_2, h^r).$$

Prove that the corresponding scheme is no longer IND-CCA secure.

Exercise 35 *Prove that Assumption 5.1 is equivalent to the standard decisional Diffie–Hellman assumption, choosing the group generator uniformly at random (see Assumption 4.1).*

Exercise 36 *Let $N = pq$ with $p \neq q$ be prime numbers as used in Assumption 5.2. Consider the group isomorphism*

$$f : \mathbb{Z}_N \times \mathbb{Z}_N^* \longrightarrow \mathbb{Z}_{N^2}^*$$
$$(a, b) \longmapsto (1 + N)^a b^N \bmod N^2.$$

Characterize the subgroup P as defined in Assumption 5.2; how large is P in comparison to $\mathbb{Z}_N^*$?

Exercise 37 *Show that the decomposition of $\mathbb{Z}_{N^2}^*$ into a direct product as claimed in Equation (5.4) indeed exists.*

Exercise 38 *Let us consider the vector space $X = \mathbb{F}_q^n$, with q prime, and $\{\alpha_1, \ldots, \alpha_n\}$ being an $\mathbb{F}_q$-vector-space basis of X. Moreover, denote by H a subgroup of the general linear group $\mathrm{GL}_n(\mathbb{F}_q)$, leaving a d-dimensional subspace L of X invariant.*

Using this setting as the starting point, describe a group action system and sketch an argument for this group action system being p-diverse.

Exercise 39 *Consider the subset membership problem from Example 5.3.*

(a) How would you describe algorithms for sampling according to the distribution I_ℓ and for selecting uniformly at random, on input an instance i, an element $x \in L$ and a witness $w \in W$ for it?

(b) The distribution I_ℓ makes use of a uniformly chosen ℓ-bit prime p. Suppose we accept some deviation from the uniform distribution on all ℓ-bit primes. Can this be exploited to sample ℓ-bit primes more efficiently?

Chapter 6

Public-key encryption using infinite groups

In the constructions we have discussed so far, the involved groups have always been finite. Thinking of tasks like sampling a random element or bounding the size of a ciphertext in advance, dealing entirely with finite sets certainly has advantages. Notwithstanding this, from the mathematical point of view, looking at a wider class of algebraic structures appears natural when looking for hard algorithmic questions that ultimately may lead to public-key encryption schemes. In infinite groups one quickly encounters questions that cannot be solved algorithmically at all. Thus, it is not surprising that a number of authors suggested conceptual public-key encryption schemes, which involve infinite groups.

In this chapter we discuss a number of these proposals, along with problems that have been encountered. The task of exploiting algorithmic problems in infinite groups to realize public-key encryption with provable security guarantees turns out to be a challenging one.

6.1 The word problem in finitely presented groups

As in the case for finite groups, we have to pay attention to how the groups we are working with are represented. The schemes discussed in the sequel are formulated in the language of *combinatorial group theory*. Roughly speaking, this area of mathematics is concerned with the questions that arise when describing a group in terms of a set of *generators* (which play the role of an alphabet) and *relators*, which play the role of rewriting rules. The language induced by such a presentation may, to some extent, be ambiguous, and several interesting problems arise when trying to relate the combinatorial properties of the language with its underlying group structure.

DEFINITION 6.1 [Free group]   *Let X be a set. Then a finite non-empty word $w \in (X \cup X^{-1})^*$ over the alphabet $X \cup X^{-1}$ is said to be* reduced, *if it does not contain any bigram xx^{-1} or $x^{-1}x$ with $x \in X$. The group* *generated by X is* free *if every reduced word in $(X \cup X^{-1})^+$ is different from the identity. If this is the case, then we say that the group is* freely generated *by X and denote this group by $F(X)$.*

By construction, a free group $F(X)$ may seem to be a rather trivial object—its elements are reduced words over an alphabet X, and the group operation is just concatenation followed by reducing modulo the trivial relations $xx^{-1} = \epsilon$ and $x^{-1}x = \epsilon$ ($x \in X$). On closer inspection, free groups turn out to have some quite remarkable (not so obvious) properties, however. In particular, every group can be obtained as quotient of a free group. The latter observation is the starting point for putting the notion of *presentation of a group* on more formal grounds.

DEFINITION 6.2 [Presentation of a group]   *Let G be a group, $X \subseteq G$ a subset of G, and R a set of finite words over the alphabet $X \cup X^{-1}$ (i. e., $R \subseteq (X \cup X^{-1})^*$). We say that (X, R) is a* presentation *of the group G whenever G is isomorphic to the quotient of a group freely generated by X modulo the normal subgroup generated by R. The elements from X are usually referred to as* generators *and the ones from R as* relators.
*   If a group G has a presentation (X, R) as above, with both X and R being finite, then G is said to be* finitely presented.

The *word problem* for a finite presentation (X, R) of group G refers to the task of finding an algorithm that can decide whether two given finite words in $X \cup X^{-1}$ represent the same group element in G. Equivalently, to solve the problem one should be able to tell whether a given word is representing the identity element of the group G.

REMARK 6.1   The careful reader may point out that this characterization of the word problem is not sufficiently precise, as it is not clear how the words presented to the algorithm are chosen, which success probability and which running time an algorithm to solve the word problem is expected to have. In fact, it is not uncommon to impose that an algorithm for solving the word problem can handle *every* possible pair of words over the alphabet $X \cup X^{-1}$, *always* succeeds in finite time, but has no restriction on the running time needed to do so. Formalizing the word problem along this line, it may not be that surprising that obtaining a proof for a property like IND-CPA security from a "hard word problem" alone is not a straightforward exercise. ⬜

6.1.1   The encryption scheme of Wagner and Magyarik

One of the first attempts to exploit the hardness of a group-theoretic problem for building an asymmetric encryption scheme dates back several decades.

At CRYPTO '84 Wagner and Magyarik suggested the use of the above-mentioned word problem in finitely presented groups as a starting point to build an encryption scheme [WM85]. Similar ideas gave rise to other encryption schemes, on which we will comment later in this section. In particular, we will see that the Polly Cracker proposal by Fellows and Koblitz [FK94] can be viewed as a proposal within this framework.

Wagner and Magyarik's encryption scheme was introduced in a rather conceptual, generic way as depicted in Figure 6.1.

$\mathcal{K}$: on input a security parameter, produces:

- a finite presentation (X, R) of a group G for which the word problem is considered to be hard;

- a subset $S \subseteq (X \cup X^{-1})^*$ such that for the quotient group $\widetilde{G}$, specified by $(X, R \cup S)$, the word problem can be solved efficiently by means of an algorithm $\mathcal{S}$;

- a set $W(\Sigma) = \{w_\sigma | \sigma \in \Sigma\}$, contained in $(X \cup X^{-1})^*$ such that if $\sigma \neq \tau$, then w_σ and w_τ are neither equivalent over G nor over $\widetilde{G}$. At this, Σ is the (finite) plaintext space.[a]

The public key consists of the presentation (X, R) and the set $W(\Sigma)$; the set S (specifying the quotient $\widetilde{G}$) constitutes the private key.

$\mathcal{E}$: on input the public key and a plaintext symbol $\sigma \in \Sigma$, sets $w = w_\sigma$ and rewrites this word w using the public relations specified by R, yielding the ciphertext.

$\mathcal{D}$: recovers the plaintext corresponding to an input word w by running $\mathcal{S}$ with the input pairs (w, w_σ) for all $\sigma \in \Sigma$.

---

[a]One can think of Σ as plaintext *alphabet* with words over Σ being encrypted letter by letter (cf. Exercise 27).

FIGURE 6.1: Wagner and Magyarik encryption scheme—generic proposal.

As the authors pointed out, some care must be taken in the key generation procedure in order to avoid very simple attacks to the above scheme: clearly, it must be the case that for most quotient groups of G with easy word problems all the words in $W(\Sigma)$ collide, i.e., represent the same group element. Further, more subtle problems may arise when turning this abstract description into a concrete encryption scheme. A number of authors analyzed the security of Wagner and Magyarik's proposal, including Levy-dit-Vehel and Perret [dVP06] and Birget et al. [BMS06]. The attack we discuss in this section was published about 20 years after Wagner and Magyarik published their scheme [GVS04] and can be categorized as a *reaction attack*, where an adversary exploits information about the validity of a ciphertext. To explain this attack, let us look at a more specific example of the construction in Figure 6.1.

Example 6.1

The key generation algorithm $\mathcal{K}$ chooses a group G given by a finite set of generators $X = \{x_1, \ldots, x_n\}$ subject to relations of three types:

(R1) $x_i x_j x_k x_l = x_l x_j x_k x_i$ $(x_i, x_j, x_k, x_l \in X \cup X^{-1})$

(R2) $x_i x_j x_k = x_k x_j x_i$ $(x_i, x_j, x_k \in X \cup X^{-1})$

(R3) $x_i x_j x_k = x_j x_k x_i$ $(x_i, x_j, x_k \in X \cup X^{-1})$

These relations can all be made trivial imposing a set of relations S of the types:

(S1) $x = \epsilon$ $(x \in X)$

(S2) $x_i = x_j$ $(x_i, x_j \in X \cup X^{-1})$

(S3) $x_i x_j = x_j x_i$ $(x_i, x_j \in X)$

By adding such a set of relations S, a quotient group $\widetilde{G}$ is built which has a presentation formed by a subset of X and a set of commutativity relations. There is a polynomial time algorithm for solving the word problem for such a group, and thus for decrypting.

Now, the set of public words $W(\Sigma)$ is constructed in such a way that most sets of relations of the mentioned types which make the public relations trivial, also force that all words in $W(\Sigma)$ become equivalent to the empty word. The authors of [WM85] suggest that for that purpose, the designer of the cryptosystem may select a (small) set $\mathcal{P}$ of non-commuting pairs, such that in each quotient of G for which any pair in $\mathcal{P}$ commutes the words in $W(\Sigma)$ vanish to the empty word ϵ, while if any other pair of generators commutes those words remain inequivalent. ⬚

Como pasa en Ejemplo 6.1

To mount a reaction attack against this scheme, we assume that the adversary has access to a *validity checking oracle* $\mathcal{O}$ which may be seen as a severely limited decryption oracle: given a candidate ciphertext—in our case a word $w \in (X \cup X^{-1})^*$—we have $\mathcal{O}(w) = 1$ if w corresponds to a valid ciphertext, (i.e., if in $\widetilde{G}$ the word w is equivalent to a w_σ for some $\sigma \in \Sigma$), and we have $\mathcal{O}(w) = 0$ otherwise (i.e., if decryption fails). We also make the following assumptions:

- there is an element $0 \in \Sigma$ such that for all $\sigma, \tau \in \Sigma \setminus \{0\}$ the words $w_0 w_\sigma$ and $w_\tau w_0$ are inequivalent in both G and $\widetilde{G}$. For a plaintext alphabet Σ of moderate size (e.g., $\Sigma = \{0, 1\}$), the attacker can simply exhaust all possible choices of w_0, and we will assume that w_0 is known to the adversary.

- The adversary knows a subset $A \subseteq (X \cup X^{-1})^*$ such that an exhaustive search over A is feasible, and from its subset

$$\bar{S} = \{a \in A : a \text{ is equivalent to } \epsilon \text{ in } \widetilde{G}\}$$

one can derive a presentation of $\widetilde{G}$ or of another quotient of G that also provides a valid private key—see [WM85, Section 4.2, Attack (b)]. In Example 6.1, the special shape (S1), (S2), (S3) of the secret relations provides the adversary with such a set A.

To find $\bar{S}$, for each $a \in A$ the adversary submits no more than two queries to the validity checking oracle $\mathcal{O}$ to decide whether $a \in \bar{S}$:

(i) The word $a w_0$ is queried to the validity checking oracle:

- If $\mathcal{O}(a w_0) = 0$, then obviously $a \notin \bar{S}$.

- If $\mathcal{O}(a w_0) = 1$, then $a \in \bar{S}$, or in $\widetilde{G}$ the word $a w_0$ is equivalent to some w_σ with $\sigma \in \Sigma \setminus \{0\}$ (and hence $a \notin \bar{S}$). To distinguish these cases, a second query can be used:

(ii) The word $w_0 a$ is queried to the validity checking oracle:

- If $\mathcal{O}(w_0 a) = 0$, then obviously $a \notin \bar{S}$.

- If $\mathcal{O}(w_0 a) = 1$, then $a \in \bar{S}$, or in $\widetilde{G}$ the word $w_0 a$ is equivalent to w_τ for some $\tau \in \Sigma \setminus \{0\}$ (and hence $a \notin \bar{S}$). In the latter case ($a \notin \bar{S}$) we conclude that $w_0 a w_0$ and $w_\tau w_0$ represent the same group element. But from the previous query we know that the situation $a \notin \bar{S}$ occurs only if $a w_0$ is equivalent to w_σ for some $\sigma \in \Sigma \setminus \{0\}$, i.e., if $w_0 a w_0$ and $w_0 w_\sigma$ represent the same group element for some $\sigma \in \Sigma \setminus \{0\}$—in contradiction to $w_0 w_\sigma$ not being equivalent to $w_\tau w_0$ for $\sigma, \tau \in \Sigma \setminus \{0\}$. In summary, the situation $\mathcal{O}(w_0 a) = 1$ and $a \notin \bar{S}$ is impossible, and $\mathcal{O}(w_0 a) = 1$ implies $a \in \bar{S}$.

Let us illustrate this attack with the above Example 6.1. In correspondence with the three types of relations (S1), (S2), and (S3) we apply the above procedure three times. First we look for relations of type (S1) by searching through the set

$$A_1 = X$$

of size n. This yields a subset $\bar{S}_1$ of A_1 with words (actually generators) that vanish in $\widetilde{G}$, and we denote by $X_2 = X \setminus \bar{S}_1$ the set of remaining "non-vanishing" generators. Next, we search through the set

$$A_2 = \{x_i x_j^{-1} : x_i \neq x_j \text{ and } x_i, x_j \in X_2 \cup X_2^{-1}\}$$

of size $O(n^2)$ to identify relations of type (S2). This yields another set $\bar{S}_2$ of words vanishing in $\widetilde{G}$, and when looking for relations of type (S3) we can restrict our attention to words in those generators which have not been identified as superfluous so far; we denote this subset of X_2 by X_3. Then the final exhaustive search covers the set

$$A_3 = \{x_i x_j x_i^{-1} x_j^{-1} : x_i \neq x_j \text{ and } x_i, x_j \in X_3\}$$

of size $O(n^2)$ and yields a set $\bar{S}_3$ of words vanishing in $\widetilde{G}$. Now the desired set $\bar{S}$ is obtained as $\bar{S} = \bar{S}_1 \cup \bar{S}_2 \cup \bar{S}_3$. Namely, $(X, R \cup \bar{S})$ is a presentation of the secret quotient $\widetilde{G}$ (where $G = (X, R)$).

6.1.2   Polly Cracker

Something similar as in the example above occurred with a polynomial-based encryption scheme that was published about 10 years after Wagner and Magyarik's proposal: the Polly Cracker scheme introduced by Fellows and Koblitz [FK94] turned out to be susceptible to a very simple chosen ciphertext attack [SG02]. Polly Cracker can essentially be taken for a special case of Wagner and Magyarik's scheme, although the former has originally been formulated in terms of multivariate polynomials and ideals in polynomial rings and not in the language of combinatorial group theory. In the reformulation from [GVS06a], however, the cryptanalysis from [SG02] follows naturally from the reaction attack we just described. Consider the reformulation of the generic Polly Cracker scheme from [FK94] given in Figure 6.2.

Once we have expressed the Polly Cracker scheme as an instantiation of Wagner and Magyarik's proposal, let us try to apply the attack described in the previous section against Polly Cracker: The secret relations have the form $x_i - \xi_i$, and if the finite field $\mathbb{F}_q$ is small, then exhausting all of these relations is certainly possible. The strategy discussed suggests to fix some $w_\sigma \in W(\Sigma)$ and a ciphertext $w = w_\sigma + (x_i - \eta)$ with η a guess for ξ_i. The problem we encounter, however, is that Polly Cracker as described *never* rejects a polynomial. So a validity checking oracle is not particulary helpful in this case, and an attack model where the adversary can access a *decryption* oracle seems natural. In our case, the plaintext corresponding to w is $w_\sigma + \xi_i - \eta$, i.e., the attacker learns ξ_i. Clearly, the choice of w_σ and η is irrelevant, and this attack is basically the one described in [SG02].

$\mathcal{K}$: on input the security parameter, this algorithm picks a finite field $\mathbb{F}_q$, a point $\xi \in \mathbb{F}_q^n$, and sets $S = \{x_1 - \xi_1, \ldots, x_n - \xi_n\}$.

Then it chooses a generating set R of an ideal $I \subseteq \mathbb{F}_q[x_1, \ldots, x_n]$ contained in the ideal spanned by S in $\mathbb{F}_q[x_1, \ldots, x_n]$.

The Abelian group $G = \mathbb{F}_q[x_1, \ldots, x_n]/(R)$ is made public, whereas

$$\widetilde{G} = \mathbb{F}_q[x_1, \ldots, x_n]/(R \cup S)$$

remains secret.

The plaintext alphabet $\Sigma = \mathbb{F}_q$ and the public set $W(\Sigma) = \Sigma$ coincide. Thus, different elements from $W(\Sigma)$ never represent the same element in G or $\widetilde{G}$.

$\mathcal{E}$: on input a plaintext $\sigma \in \Sigma$ and the public key, the encryption algorithm rewrites $w_\sigma = \sigma$ with the public relations in R, i. e., by adding an element of the ideal spanned by R.

$\mathcal{D}$: given a ciphertext word w, this algorithm subtracts an element of $W(\Sigma)$ from w and checks whether the obtained difference reduces to 0 modulo the secret (Gröbner) basis S of the ideal spanned by $R \cup S$. In other words, it checks whether the difference represents the identity in $\widetilde{G}$.[a]

---

[a]More efficiently, the ciphertext (a polynomial) can be evaluated at ξ to obtain the plaintext.

FIGURE 6.2: Polly Cracker scheme by Fellows and Koblitz.

6.1.3 A successor of the Wagner–Magyarik scheme

In [BMS06], Birget et al. present a security analysis of Wagner and Magyarik's scheme and, aiming at a more secure construction, introduce a new design building on a similar problem as a base. The proposal by Birget et al. is described in Figure 6.3. The authors include some precise instructions on how to design concrete instances, and the paper mentions possible platform groups. Nonetheless, similar to the proposal we discussed in Section 6.1.1 it is rather general.

Already from the high-level specification it can be seen that the scheme fails to meet the security goal of ciphertext indistinguishability against an adversary who has access to a decryption oracle (IND-CCA as discussed in Section 4.2). Let $\mathcal{A}$ be an adversary having access to a decryption oracle $\mathcal{O}_\mathcal{D}$ as described in Definition 4.5. For the challenge plaintexts to be found in the find-phase, $\mathcal{A}$ has no other choice than using 0 and 1, as the plaintext space contains only two elements. Let c be the challenge ciphertext received by the adversary $\mathcal{A}$. To find out if c decrypts to 0 or 1, the adversary may query $\mathcal{O}_\mathcal{D}$

$\mathcal{K}$: on input the security parameter 1^ℓ, this algorithm

- outputs a finite presentation (X, R) of a group G which right-acts faithfully and transitively on $\{0, 1, 2\}^*$. For $g \in G$ and $v \in \{0, 1, 2\}^*$ we write $(v)g$ for the result of the action of g on v.

- chooses uniformly at random three words $x, z^0, z^1 \in \{0, 1, 2\}^*$ of length between ℓ and 2ℓ.

- for each $b \in \{0, 1\}$ chooses a set of $m - 1$ *intermediary words* over $\{0, 1, 2\}$, which we denote by $\{z_1^b, \ldots, z_{m-1}^b\}$. The sets $\{z^0, z_1^0, \ldots, z_{m-1}^0\}$ and $\{z^1, z_1^1, \ldots, z_{m-1}^1\}$ are disjoint.

- chooses for each $b \in \{0, 1\}$ a *system of words*, i.e., a sequence of m finite sets $Z_1^b, \ldots, Z_m^b$. Each such set Z_i^b is a small set of words over $X \cup X^{-1}$ such that $(z_{j-1}^b)w = z_j^b$ for all $w \in Z_j^b$ and $j = 2, \ldots, m - 1$.

  Also, for each $w \in Z_1^b$, $(x)w = z_1^b$ and for each $w \in Z_m^b$, $(z_{m-1}^b)w = z^b$. We can summarize this in an action diagram:

$$x \xrightarrow{Z_1^b} z_1^b \xrightarrow{Z_2^b} z_2^b \xrightarrow{Z_3^b} \cdots \xrightarrow{Z_{m-1}^b} z_{m-1}^b \xrightarrow{Z_m^b} z_b.$$

The private key is (x, z^0, z^1).

The intermediary words z_i^b, $i = 1, \ldots, m - 1$, $b = 1, 0$ are kept secret as well, but they are not needed for decryption.

The presentation (X, R) of G and the two systems of words $(Z_1^0, \ldots, Z_m^0)$ and $(Z_1^1, \ldots, Z_m^1)$ constitute the public key.

$\mathcal{E}$: on input a plaintext bit b, for each $j = 1, \ldots, m$ a group element $g_j \in Z_j^b$ is chosen uniformly at random, and the word $g_1 g_2 \ldots g_m$ is rewritten using the public relators in R. The word resulting from this rewriting process is the ciphertext.

$\mathcal{D}$: given a word $c \in (X \cup X^{-1})^*$, the action of c on (x) is computed. If the result of this action is equal to z^b, the plaintext b is output.

FIGURE 6.3:   Encryption scheme by Birget et al.

with any word in $(X \cup X^{-1})^* \setminus \{c\}$. Thus, $\mathcal{A}$ can simply choose two different words $w, \hat{w} \in Z_m^0$ and submit the word $\hat{w}w^{-1}c$ to $\mathcal{O}_\mathcal{D}$. Then $\mathcal{O}_\mathcal{D}$ outputs the correct decryption b of c:

$$x \xrightarrow{\hat{w}} z_1^0 \xrightarrow{w^{-1}} x \xrightarrow{c} z^b.$$

To avoid such a chosen ciphertext attack, one could try to first establish the exact hardness assumption needed to prove security against chosen plaintext attacks only. Once a plausible candidate group for such a setting is found and the details of the scheme in Figure 6.3 could be filled in appropriately, one could try to use a general technique to lift an IND-CPA secure public-key encryption scheme to an IND-CCA secure one. This road does not seem to be a trivial one, however, one problem already being to identify a hardness assumption that can be used with confidence, i.e., which is not only precisely stated but also "natural enough" from a mathematical point of view. Note for example that in the above description, it is not exactly the word problem in the underlying group that forms the starting point of the scheme. Rather, a *promise problem* is used: given a word $w \in (X \cup X^{-1})^*$, which is known (promised) to be equivalent to either a word in $Z_1^1 \cdots Z_m^1$ or to a word in $Z_1^0 \cdots Z_m^0$, decide which of the two is actually the case.

6.2 Using a group that is not finitely presentable?

Invoking seemingly "more complex" groups as a mathematical platform in the hope of increasing the security level of a cryptographic construction is not always a good strategy. To illustrate this, let us revisit a proposal by Garzon and Zalcstein [GZ91], which uses a similar starting point as Wagner and Magyarik's construction. Differing from the latter, however, in Garzon and Zalcstein's proposal the secret quotient group is *not* finitely presented—notwithstanding this, there is an efficient algorithm solving its word problem. Grigorchuk groups are known for their interesting theoretic properties, and we refer the interested reader to Garzon and Zalcstein's paper [GZ91] for a more detailed description of the encryption scheme we discuss below, and for more details on Grigorchuk groups, including a discussion of how to interpret elements of such a group as permutations of the infinite complete binary tree. For our purposes, a brief outline of the scheme is enough, and we then describe a rather simple, but effective, attack as discussed in [GVHMS04].

Given an infinite ternary sequence $\chi = (\chi_i)_{i \geq 1}$ over the alphabet $\{0, 1, 2\}$, the *Grigorchuk group* G_χ associated with χ is an infinite 4-generator group generated by four involutions a, b_χ, c_χ, and d_χ—as indicated by the notation, the latter three depend on the sequence χ. If at least two of the symbols 0, 1, and 2 repeat infinitely often in the sequence χ, then the group G_χ is not finitely presentable and has a property known as *subexponential growth*. The latter property refers to the number of group elements that can be expressed in the generators as words of a given length. While not being essential for our cryptanalytic discussion, from a group-theoretic point of view, this property of Grigorchuk groups is quite prominent.

For the values of χ considered throughout, one can check efficiently whether a word $w \in \{a, b_\chi, c_\chi, d_\chi\}^*$ represents the identity in G_χ, if the first $\lceil \log_2 |w| \rceil$ positions of χ are known, where $|\cdot|$ denotes the word length (i.e., the number of generators appearing in w). So the word problem in G_χ can be solved in polynomial time given a sufficiently large fragment of χ. Making use of this fact, the Garzon–Zalcstein public-key encryption scheme to encrypt one plaintext bit can be outlined as shown in Figure 6.4.

$\mathcal{K}$: chooses an infinite ternary sequence χ such that both G_χ is not finitely presentable and the word problem in G_χ can be solved efficiently.

The private key is a (description of a) *Turing machine* M_{G_χ} solving the word problem in the Grigorchuk group G_χ.[a]

Further, a finite subset of relators R of G_χ is chosen along with two words w_0, w_1 in the generators $a, b_\chi, c_\chi, d_\chi$ representing distinct group elements $\omega_0, \omega_1 \in G_\chi$. The size of R, as well as $|w_0|$ and $|w_1|$ should be polynomial in the security parameter.

The public key is formed by R, w_0 and w_1.

$\mathcal{E}$: on input a plaintext bit $b \in \{0, 1\}$, this algorithm rewrites the word w_b by applying repeatedly some of the public relators R, yielding a ciphertext $c \in \{a, b_\chi, c_\chi, d_\chi\}^*$. At this, $|c|$ is required to be of length polynomial in the security parameter.

$\mathcal{D}$: to decrypt a ciphertext $c \in \{a, b_\chi, c_\chi, d_\chi\}^*$, this algorithm uses the secret Turing machine M_{G_χ} for checking whether in G_χ the group element represented by c is equal to ω_0 or ω_1. Accordingly, the plaintext 0 or 1 is output.

---

[a] M_{G_χ} should be polynomial in the security parameter.

FIGURE 6.4: Garzon–Zalcstein encryption scheme—generic proposal.

Obviously the above description is not very detailed and there are several open questions one should answer when designing an implementation. On the efficiency side, we observe that the encryption algorithm handles only one bit at a time, so depending on the details of the rewriting step, the message expansion might be substantial. Actually, from the above description it is not clear how exactly the rewriting process in the encryption step should be carried out. Still, already at this conceptual level of the proposal something can be said about the security properties of the scheme.

Garzon and Zalcstein's paper [GZ91] adopts the assumption that verifying the correctness of a guess for the first few positions of χ is difficult, because of Alice's public key revealing only the finite subset R of relators of the (not finitely presentable) group G_χ. Notwithstanding this property of G_χ, during encryption only the finite public set of relators R is used. As noted in [GVHMS04], an adversary can exploit this to derive an alternative secret key. Let $N = \max\{|r| : r \in R \cup \{w_0^{-1}w_1\}\}$ be the maximum of the lengths of the "relational words" in Alice's public key and of the word $w_0^{-1}w_1$, where w_0, w_1 are Alice's two distinct public words. Then, having in mind a sensible size of the public key, one should assume that the number $B = \lceil \log_2(N) \rceil$ is rather small and that a brute-force search over all 3^B words in $\{0, 1, 2\}^B$ is feasible. If we denote by $\bar{\chi}$ the finite sequence consisting of the first B elements of the secret χ, then an adversary can use the following strategy:

1. For each possible choice of the first B positions of χ, check if both

   - each element of R vanishes and
   - $w_0^{-1}w_1$ does not vanish in G_χ.

   By construction, at least one possible choice must satisfy these conditions (as in particular $\bar{\chi}$ does).

2. Once such a finite sequence is found, complete it in an arbitrary suitable manner (e. g., by appending zeroes) to an infinite ternary sequence χ'.

The public key depends on the finite sequence $\bar{\chi}$ only, i. e., the public key could have been derived from χ' as well as from χ. Thus, a Turing maching solving the word problem in $G_{\chi'}$ now correctly decrypts all ciphertexts encrypted under Alice's public key.

Several other encryption schemes using word or rewriting problems in groups presented with generators and relations (or in weaker structures, such as monoids) can be found in the literature—see, for instance [MS90, STAS02, ATS03, BMS06, SZ09]. Many of them have been cryptanalyzed using reaction attacks, simple chosen-ciphertext attacks or even—as in the previous example—a weaker attack model [GVS01, GVS06b, GVP07, GV05, GVT06]. While cryptographically significant, from a group-theoretic point of view such attacks are not always interesting—no mathematical insight is immediately implied "just" because an adversary can find a plaintext or distinguish the encryptions of two plaintexts. Here the use of provable security guarantees offers an attractive feature: if, say, IND-CCA security of a scheme provably reduces to an interesting group-theoretic problem, then a successful attack within the used security model immediately yields an algorithm to solve an interesting problem. In the next section we want to take a look at an area of group-theoretic cryptography which flourished at the beginning of this century and where the exchange between the cryptography and the group theory research communities has been quite fruitful, even though in the end no cryptographic scheme emerged from it whose security is widely accepted by the research community.

6.3 Braid groups in cryptography

Let us take a look at *braids* as first introduced by Emil Artin in [Art47]. They appear in different areas of mathematics and form the mathematical basis for a number of cryptographic proposals that were made around the beginning of the 21st century. The schemes are mathematically very appealing and the constructions likely to be efficient enough to be practical. So it is not suprising that these proposals attracted a lot of attention, both from the group theory and the cryptographic community. Unfortunately, a number of practical problems arose after the proposals underwent more careful scrutiny. Still, the elegant ideas behind these constructions continue to stimulate research in the area. Two surveys on the topic have been written by P. Dehornoy [Deh04] and D. Garber [Gar07], and much of what we explore in this section follows their presentations.

6.3.1 Basics on braid groups

Braid groups admit several equivalent definitions. Here we use the so-called *Artin presentation* as a starting point:

DEFINITION 6.3 [Braid group: Artin presentation]   *For any natural number $n \geq 2$, the* braid group on n strands, B_n, *is defined by the presentation (X, R), with generator set $X = \{\sigma_1, \ldots, \sigma_{n-1}\}$, and relator set*

$$R = \{\sigma_i \sigma_j (\sigma_j \sigma_i)^{-1} \text{ for } 1 \leq i, j \leq n-1, |i-j| \geq 2\} \quad \cup$$
$$\{\sigma_i \sigma_j \sigma_i (\sigma_j \sigma_i \sigma_j)^{-1} \text{ for } 1 \leq i, j \leq n-1, |i-j| = 1\}.$$

The generators σ_i, $i = 1, \ldots, n-1$ are referred to as Artin generators, *and the elements of B_n are called n-braids.*

This presentation affords an intuitive geometric interpretation: suppose we have n strands that are linked to a solid bar at the ceiling, hanging down to the floor, where they are linked to another solid bar, parallel to the first

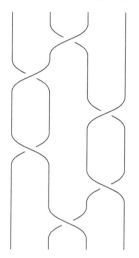

FIGURE 6.5: Geometric interpretation of $\sigma_2\sigma_1\sigma_3\sigma_1^{-1}\sigma_3^{-1}\sigma_2^{-1}$.

one. If we number the strands consecutively from 1 to n, the generator σ_i represents a crossing of the ith strand under the $i + 1$-st strand, and the group operation can be interpreted as concatenating two braids. The identity, which we denote by $\epsilon \in B_n$, corresponds to a braid with no crossings at all, and the relations in Definition 6.3 reflect the commutativity of certain crossing operations. Figure 6.5 shows a geometric interpretation of the 3-braid $\sigma_2\sigma_1\sigma_3\sigma_1^{-1}\sigma_3^{-1}\sigma_2^{-1}$. This braid word represents the identity, and in the diagram one can see how the relation $\sigma_1\sigma_3 = \sigma_3\sigma_1$ allows to undo the crossings. For a more detailed discussion of the geometric interpretation of braids, the interested reader is encouraged to consult a textbook about braid groups such as [MK99].

To understand the braid groups B_n, it is helpful to look at their relation with the symmetric groups S_n. For this, we define a group homomorphism

$$\begin{aligned}\phi : B_n &\longrightarrow S_n \\ \sigma_i &\longmapsto (i, i+1) \quad (i = 1, \ldots, n)\end{aligned} \tag{6.1}$$

That is, ϕ maps each Artin generator to the corresponding transposition of indices of adjacent strands. It is easy to see that ϕ defines an epimorphism, and for cryptanalytic purposes one can hope that this homomorphism helps to translate algorithmic problems in the infinite group B_n into algorithmic problems in the finite group S_n. Of course, ϕ is not one-to-one, and its non-trivial kernel $\ker(\phi)$ is commonly known as the subgroup of *pure braids*.

Similarly as for Artin generators, the so-called *band generators* or *Birman–Ko–Lee generators* [BKL98]—named after Joan Birman, Ki Hyoung Ko, and Sang Jin Lee—are elements that can be interpreted as braids in which exactly

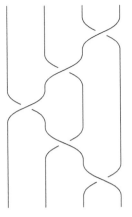

FIGURE 6.6:   Birman–Ko–Lee generator $a_{41} = (\sigma_3\sigma_2)\sigma_1(\sigma_3\sigma_2)^{-1}$.

two strands are "swapped." Differing from the Artin presentation, however, now we do not restrict to generators whose ϕ-image is a transposition of the form $(i, i+1)$. Instead, *each* of the $\binom{n}{2}$ transpositions in S_n has a ϕ-preimage among the band generators:

DEFINITION 6.4 [Braid group: Birman–Ko–Lee generators]   *For each $1 \le s < t \le n$ consider the braid*

$$a_{ts} = (\sigma_{t-1}\sigma_{t-2}\cdots\sigma_{s+1})\sigma_s(\sigma_{t-1}\sigma_{t-2}\cdots\sigma_{s+1})^{-1} \in B_n.$$

Such a braid is referred to as a band generator *or* Birman–Ko–Lee generator *of B_n.*

As an example, Figure 6.6 gives the geometric interpretation of the band generator $a_{41} \in B_4$, expressed as a word in Artin generators: this generator swaps the first and fourth strand and leaves all other strands unchanged. The swap takes place "in front" of the other strands. As shown in [BKL98], we have the relations

$$a_{ts}a_{rq} = a_{rq}a_{ts} \qquad \text{if } (t-r)(t-q)(s-r)(s-q) > 0$$
$$a_{ts}a_{sr} = a_{tr}a_{ts} = a_{sr}a_{tr} \text{ for all } r, s, t \text{ with } 1 \le r < s < t \le n\text{'}$$

yielding another presentation of B_n in terms of generators and relations. A main motivation for considering such an alternative presentation of the braid group is a computational one—being able to perform certain computations in B_n more efficiently. For the implementor of a cryptographic scheme, such efficiency considerations can be crucial, and in the next section we look a bit more closely at some computational tasks in braid groups.

6.3.2   Some computational problems in the braid group B_n

One of the reasons for the appeal of braid goups as a candidate platform for cryptography lies in the availability of convenient *normal forms*—it is possible to compute efficiently a unique representation of a group element. Such a property may seem a triviality when thinking of common representations of multiplicative subgroups of a finite field or of subgroups of an elliptic curve, but in the context of finitely presented groups this is a remarkable feature. In particular, a normal form that can be computed efficiently enables an efficient solution of the word problem, and we can identify group elements with their normal form.

The word problem in B_n.   Different normal forms have been explored in the literature, and what we discuss here is known as *Garside normal form*—named after Frank A. Garside—or as *greedy normal form*. More information on normal forms for braid groups can be found, for instance, in work by Epstein et al. [ECH$^+$92] and Birman et al. [BKL98].

By B_n^+ we denote the monoid of *positive braids*, i. e., the *monoid* generated by the Artin generators $\sigma_1, \ldots, \sigma_{n-1}$. In other words, B_n^+ consists of all braids on n strands that can be expressed as a finite word in the Artin generators without involving any inverses σ_i^{-1}. An important example of a positive braid is the *fundamental braid*, which plays a central role in the computation of the greedy normal form.

DEFINITION 6.5 [Fundamental braid]
The fundamental braid $\Delta_n \in B_n$, *is defined as*

$$\Delta_n = (\sigma_1 \ldots \sigma_{n-1})(\sigma_1 \ldots \sigma_{n-2}) \ldots \sigma_1.$$

As a concrete example, Figure 6.7 shows the geometric interpretation of the fundamental braid Δ_4.

A second core ingredient for the Garside normal form are so-called *permutation braids*—restricting the above-mentioned epimorphism $\phi : B_n \longrightarrow S_n$ to permutation braids results in a bijection of sets. In particular, there are exactly $n!$ permutation braids, and these are in one-to-one correspondence with the permutations of $1, \ldots, n$. To give a precise definition, it is helpful to introduce some terminology: we call a braid $b_2 \in B_n$ a *prefix* of a braid $b_1 \in B_n$ and write $b_2 \preceq b_1$ if there exists a positive braid $b_3 \in B_n^+$ such that $b_1 = b_2 b_3$.

DEFINITION 6.6 [Permutation braid]   *A braid $b \in B_n$ is called a* permutation braid *or a* simple braid *if the relation $\epsilon \preceq b \preceq \Delta_n$ holds.*

With each permutation n-braid p, we associate a *starting set* that contains

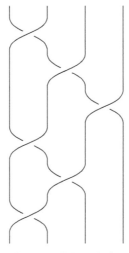

FIGURE 6.7: Fundamental braid $\Delta_4 = (\sigma_1\sigma_2\sigma_3)(\sigma_1\sigma_2)\sigma_1$.

all indices i such that $\sigma_i \preceq p$:

$$S(p) = \{i \in \{1,\ldots,n-1\} : \exists p' \in B_n^+ : p = \sigma_i p'\}$$

Similarly, we define a *finishing set*

$$F(p) = \{i \in \{1,\ldots,n-1\} : \exists p' \in B_n^+ : p = p'\sigma_i\}.$$

Example 6.2 [Fundamental braid]
For the fundamental braid Δ_n the starting and finishing sets are identical and equal to $\{1,\ldots,n-1\}$. ☐

With this, we are ready to describe the Garside normal form of a braid:

THEOREM 6.1 [Garside normal form]
For each braid $b \in B_n$ there is a unique representation $b = \Delta_n^r p_1 \cdots p_k$, such that

- *the integer $r \in \mathbb{Z}$ is maximal,*

- *$p_1,\ldots,p_k$ are permutation braids,*

- *$p_k \neq \epsilon$, and*

- *$S(p_{i+1}) \subseteq F(p_i)$ for $i = 1,\ldots,k-1$.*

This representation is called Garside normal form *or* greedy normal form *of* b.

We do not prove this theorem here, but as discussed in [ECH+92], an efficient rewriting procedure allows to compute the Garside normal form of a braid b in time $O(|w|^2 \cdot n \log(n))$ with $|w|$ denoting the length of a finite braid word w representing b in the Artin generators. Deriving the greedy normal form involves three main steps:

1. Any negative generator σ_i^{-1} is replaced by $\Delta_n^{-1} p_i$, with a permutation braid p_i.

2. Exploiting the relations $\sigma_i \Delta_n^{-1} = \Delta_n^{-1} \sigma_{n-i}$, all occurring fundamental braids Δ_n are moved to the left of the braid word.

3. Steps 1 and 2 yield a representation $b = \Delta_n^k b'$ with $b' \in B_n^+$. The positive braid b' is decomposed into simple braids such that the condition on starting and finishing sets in Theorem 6.1 is fulfilled.

Example 6.3
To find the Garside normal form of

$$b = \sigma_2 \sigma_3^{-1} \sigma_3^{-1} \sigma_2 \in B_4,$$

we replace both occurrences of σ_3^{-1} by $\Delta_4^{-1} \sigma_3 \sigma_2 \sigma_1 \sigma_3 \sigma_2$ and obtain

$$b = \sigma_2 (\Delta_4^{-1} \sigma_3 \sigma_2 \sigma_1 \sigma_3 \sigma_2)(\Delta_4^{-1} \sigma_3 \sigma_2 \sigma_1 \sigma_3 \sigma_2) \sigma_2.$$

Moving the first Δ_4^{-1} to the left, we obtain

$$b = \Delta_4^{-1} \sigma_2 \sigma_3 \sigma_2 \sigma_1 \sigma_3 \sigma_2 (\Delta_4^{-1} \sigma_3 \sigma_2 \sigma_1 \sigma_3 \sigma_2) \sigma_2.$$

Proceeding analogously with the second occurrence of Δ_4^{-1} yields

$$b = \Delta_4^{-2} \sigma_2 \sigma_1 \sigma_2 \sigma_3 \sigma_1 \sigma_2 \sigma_3 \sigma_2 \sigma_1 \sigma_3 \sigma_2 \sigma_2.$$

Finally, expressing $\sigma_2 \sigma_1 \sigma_2 \sigma_3 \sigma_1 \sigma_2 \sigma_3 \sigma_2 \sigma_1 \sigma_3 \sigma_2 \sigma_2$ as a product of permutation braids results in the Garside normal form

$$b = \Delta_4^{-2} \sigma_2 (\sigma_2 \sigma_1 \sigma_3 \sigma_2 \sigma_3)(\sigma_3 \sigma_2 \sigma_1 \sigma_3 \sigma_2) \sigma_2.$$

☐

For the Birman–Ko–Lee presentation, another normal form exists which can be derived in time $O(|w|^2 \cdot n)$, and it deserves to be noted that not only normal forms can be used to solve the word problem in braid groups. For instance, Dehornoy presents in [Deh97] an alternative method called *handle reduction*, and Garber et al. discuss another solution to the word problem in [GKT02].

The conjugacy problem in B_n. For a finite cyclic group G with generator g of prime order p, the multiplicative group $\mathbb{F}_p^*$ acts on G through exponentiation, and this group action lies at the heart of the ElGamal encryption scheme. For cryptographic proposals related to braid groups, another group action plays a similarly crucial role: the braid group acts on itself through conjugation, and different flavors of conjugacy problems have been used in connection with cryptographic proposals. Leaving aside the technical questions that need to be addressed by a formal definition, four variants naturally come to mind:

Conjugacy decision problem: Given $a, b \in B_n$, determine whether they are conjugate, i.e., whether there exists $x \in B_n$ such that $a = x^{-1}bx$.

Conjugacy search problem: Given two braids $a, b \in B_n$, that are known to be conjugate, find a conjugating $x \in B_n$ such that $a = x^{-1}bx$.

Braid Diffie–Hellman decision problem: Given braids $a, b, c, d \in B_n$, with $b = x^{-1}ax$ and $c = y^{-1}ay$, for some commuting braids $x, y \in B_n$ (i.e., $xy = yx$), determine whether $d = (xy)^{-1}a(xy)$.

Braid Diffie–Hellman search problem: Given $a, b, c \in B_n$, such that there exist $x, y \in B_n$ with $b = x^{-1}ax$ and $c = y^{-1}ay$, with $xy = yx$, compute $d = (xy)^{-1}a(xy)$.

Unlike $\mathbb{F}_p^*$, the braid group is not Abelian, and for $g, k, x \in B_n$ the relation $(g^x)^k = (g^k)^x$, which is crucial for decryption in the ElGamal scheme, does not have to hold. To enfore a commutativity condition as used in the last two variants of the conjugacy problem, one can choose x and k from particular subgroups. For instance, we can choose them from the subgroup of left braids $LB_n \leq B_n$ and right braids $RB_n \leq B_n$, respectively, where LB_n is generated by $\sigma_1, \ldots, \sigma_{\lfloor \frac{n}{2} \rfloor - 1}$ and RB_n is generated by $\sigma_{\lfloor \frac{n}{2} \rfloor + 1}, \ldots, \sigma_{n-1}$. In this way, every element in LB_n commutes with every element in RB_n. Using the "commuting exponents" and invoking a random oracle H to map (the normal form of) a braid group element into a bitstring, one can mimic the usual ElGamal procedure as shown in Figure 6.8. Such a variant of ElGamal has been proposed at CRYPTO 2000 by Ko et al. [KLC$^+$00].

The description of the encryption scheme in Figure 6.8 lacks some important details. Specifically, the choice of the "random" braids is not obvious. Unlike for the case of finite groups here we have no obvious uniform random choice from a finite set available. Only recently, Gebhardt et al. have presented a polynomial time algorithm for generating positive braids of a given length as words in Artin generators according to a uniform distribution [GGM13].

It remains a challenging problem to solve all of the above variants of the conjugacy problem efficiently for arbitrary input choices, but from a cryptanalytic perspective already heuristic attacks with a non-negligible success probability pose a threat. Different algorithms have been proposed to solve

$\mathcal{K}$: a probabilistic *key generation* algorithm that on input a security parameter 1^ℓ chooses a braid $g \in B_{n(\ell)}$, a left braid $a \in LB_{n(\ell)}$ and computes $y = aga^{-1}$.

The public key is the pair $(g, y) \in B^2_{n(\ell)}$, and the secret key is the braid a.

$\mathcal{E}$: on input a public key $(g, y) \in B^2_{n(\ell)}$ and a message $m \in \{0, 1\}^\ell$ this algorithm chooses a random right braid $k \in RB_{n(\ell)}$ and returns the ciphertext $(kgk^{-1}, H(kyg^{-1}) \oplus m) \in B_{n(\ell)} \times \{0, 1\}^\ell$.

$\mathcal{D}$: to recover the plaintext corresponding to an input ciphertext $(h, c) \in B_{n(\ell)} \times \{0, 1\}^\ell$ with private key a, this (deterministic) algorithm computes $m = H(aha^{-1}) \oplus c$.

FIGURE 6.8: Braid-based encryption scheme with a construction by Ko et al. using a random oracle $H : B_{n(\ell)} \longrightarrow \{0, 1\}^\ell$.

conjugacy-type problems in braid groups, some of them with very high success probabilities against proposed parameter choices. Following the exposition of [Gar07], we give here a brief survey of such attacks. Many more details can be found in the papers [BGGM07a, BGGM08, BGGM07b], and from a practical point it is fair to say that at this stage no parameter selection for the above ElGamal-type scheme has received sufficient confidence from the cryptographic research community.

What is the basic idea for tackling a conjugacy decision problem or a conjugacy search problem? Typically, to each braid a a small finite subset I_a of a's conjugacy class is associated that characterizes the conjugacy class in the sense that a is conjugate to another braid b if and only $I_a = I_b$. To obtain a feasible attack strategy, it should be possible to efficiently obtain a representative $\hat{a} \in I_a$ along with a "witness" $x \in B_n$ such that $x^{-1}ax = \hat{a}$. Finally, it should be possible to enumerate all elements of I_a from a given representative $\hat{a} \in I_a$. The cost for the latter step will depend on the size of I_a, and constructing such sets with a provably small bound is a main difficulty in finding *provably* efficient solutions to the conjugacy search problem.

Independent of the theoretical bound on the size of I_a we can establish, on input two braids $a, b \in B_n$ we can try to solve the corresponding instance of the conjugacy search or decision problem as follows:

(i) Find representatives $\hat{a} \in I_A$ and $\hat{b} \in I_B$.

(ii) Enumerate elements of I_a (along with a witness) until one of the following occurs:

   (a) The braid $\hat{b}$ is found as an element of I_a proving a and b to be conjugate—and a conjugating element can be obtained from the witnesses.

(b) The entire set I_a is exhausted without finding $\hat{b}$, thereby proving that a and b are not conjugate.

Several choices for the sets I_a have been considered in the literature, carrying names like *summit set*, *super summit set*, *ultra summit set*, or *reduced supper summit set*. Qualifiers like "super" or "ultra" are commonly used to denote a reduction in size compared to previously considered sets—the simpler and smaller such sets are, the more efficient is the resulting algorithm derived from the above strategy.

Using the above technique and other sophisticated geometric techniques, J. Birman, V. Gebhardt, and J. González Meneses [BGGM07b] could identify a polynomial time algorithm for the conjugacy search problem for so-called *periodic braids*. Furthermore, the same authors have proven that the conjugacy search problem could be solved for all instances if a polynomial time algorithm for a type of braids known as *rigid braids* were found. Such results are very interesting from the perspective of computational group theory, but somewhat ambivalent from a cryptographic perspective: on the one hand, we learn about certain types of problem instances that should be avoided, while on the other hand it seems worrisome that a polynomial time solution is already available for at least one type of braids.

In addition to trying to construct a general solution, heuristic strategies have been explored to tackle instances of a conjugacy problem. These strategies include the following:

Neighboring representatives in the super summit set. Aiming at a solution of the search version of the conjugacy problem, in [HS02] a very basic version of the above general strategy was used: given two (conjugate) braids a and b, compute representatives $\hat{a}$ and $\hat{b}$, respectively, and corresponding witnesses. More specifically, the super summit set was used, where this computation can be realized efficiently. Next, an attempt is made to find a *permutation braid* conjugating one representative $\hat{a}$ to the other $\hat{b}$. Despite being far from constituting a general solution to a conjugacy search problem, this seemingly simple strategy turned out to be a very effective attack when invoked with concrete instances that had been proposed for cryptographic use in the literature.

Length-based attacks. These types of attacks were introduced by J. Hughes and A. Tannenbaum [HT03] and rely on finding a suitable "length function" L. For instance, L may be closely related to the number of generators (of a certain type) involved in some normal form representation of a braid. The idea is that if, say, an Artin generator σ_i appears last in the conjugating element x, then one may hope that the length of $\sigma_i x^{-1} a x \sigma_i^{-1}$ is smaller than that of $x^{-1}ax$. This should allow to iteratively read off a word representing the group element conjugating a to

$b = x^{-1}ax$. For a more careful discussion of this line of attack we refer to [GKT$^+$06, MU07].

Linearization attacks. There exist several different ways to linearize braid groups, i. e., to homomorphically map the braid group B_n onto a certain group of matrices. The idea of a linearization attack is to solve a conjugacy problem in a matrix group and then try to recover a preimage in B_n of the (matrix) solution found. This approach is the key idea behind the cryptanalyses in [Hug02, CJ03], which more specifically invoke the *Burau representation* and the *Lawrence–Krammer representation* of braid groups. We will learn more about this in Section 9.3, where we discuss recently proposed algorithms for violating the security of braid-based key exchange protocols.

6.4 Summary and further reading

With Wagner and Magyarik's proposal and a subsequent proposal by Birget et al. we have seen two attempts to build cryptographic schemes in the context of combinatorial group theory. Polly Cracker offers an example of a scheme that is naturally formulated in the language of polynomial rings and ideals, but that can also be interpreted in a group-theoretic framework. For each of these examples we have seen an attack, showing that getting from a computationally hard (or even undecidable) problem to a cryptographically secure scheme is a non-trivial challenge. The example with Grigorchuk groups that we have seen illustrates that the complexity of an algebraic structure as such is not an adequate indicator for cryptographic strength.

Braid groups are algorithmically quite appealing, and as we have seen, the conjugacy problem in these groups forms the basis for an ElGamal-like encryption scheme. The obvious line of attack against this scheme is to tackle the conjugacy problem, and as indicated, various approaches to this problem have been proposed, resulting in cryptanalytic successes. Outside cryptography, the theory of braid groups has a long history, and for the reader who would like to learn more about the mathematical theory of braids, a textbook by Murasugi and Kurpita [MK99] can serve as a starting point, for instance. On the cryptographic side, much of the initial optimism in using braid groups for deriving secure public-key encryption schemes has been lost, but the rich algorithmic structure of B_n remains tempting. The surveys by Dehornoy [Deh04] and Garber [Gar07] are good starting points for learning more about cryptographic uses of the braid group that have been considered. A more detailed discussion of length attacks against a (generalized) conjugacy problem can be found in [GKT$^+$06].

It also deserves to be said that the Polly Cracker cryptosystem has inspired follow-up work. In particular, there has been interest in establishing provable guarantees for a construction along this line [AFFP11, Her12]. For doing so, the so-called *Learning With Errors (LWE)* problem comes into play, an algorithmic problem that deals with the task of learning a function based on noisy samples. Interestingly, this problem has also been used by Baumslag et al. when trying to derive an encryption scheme from Burnside groups [BFN+11]. More specifically, they use a variation of this problem coined *learning homomorphisms with errors*, working with $B(n, 3)$, the free Burnside group of exponent 3. More specifically,

$$B(n, k) = (\{x_1, \ldots, x_n\}, \{w^k : w \in \{x_1, \ldots, x_k\}^*\}),$$

and for $k = 3$ this group is known to be finite. Hence, the approach suggested in [BFN+11] suggests a use of the framework of combinatorial group theory in combination with finite groups. For the reader who wants to get more acquainted with the foundations of combinatorial group theory in general, the textbook by Baumslag [Bau93] can serve as a starting point.

6.5   Exercises

Exercise 40 *Show that*
$$\langle a, b | a^2 = b^3 \rangle$$
is a presentation of the braid group B_3.

Exercise 41 *Consider the set P_n of pure braids which we defined as the kernel of the epimorphism $\phi : B_n \longrightarrow S_n$ in (6.1). Prove, using Definition 1.6, that it is indeed a normal subgroup of B_n.*

Exercise 42 *A motivation for the design of the Polly Cracker cryptosystem was that the public key can describe an instance of an NP-complete problem with the secret key encoding a solution. So for recovering the secret key from the public key alone, a (hopefully hard) instance of an NP-complete problem has to be solved. The solvability of quadratic equations over a free group [KLMT10] is an example of an NP-complete problem that is phrased in the language of combinatorial group theory.*

Here we consider an example from Fellows and Koblitz's original paper [FK94], deriving from Graph 3-Colorability. Let G be a graph with finite vertex set V and edge set E, such that V has a 3-coloring. This means there exists a function $\lambda : V \longrightarrow \{blue, green, red\}$ with $\lambda(v) \neq \lambda(v')$ for all $\{v, v'\} \in E$. Find a set of polynomials with coefficients in a finite field such that knowledge of a common zero of P is equivalent to knowing a function λ as above.

Exercise 43 *Experiment with different strategies to sample a "random" braid, aiming at the generation of hard instances of a conjugacy search problem. For instance, one could try to fix the number of strands n and then form random words of fixed length in the Artin generators. To scale the difficulty of the instances, one could try to increase the number of strands, keeping the word length fixed. Is this a good strategy?*

Exercise 44 *Describe an (efficient) algorithm that on input two permutations $\sigma, \pi \in S_n$ (in cycle notation) decides if σ and π are conjugate and if yes finds a conjugating permutation $\tau \in S_n$ such that $\sigma = \tau^{-1}\pi\tau$.*

Exercise 45 *Let $G \leq S_n$ be a permutation group. You can assume that generators of G are given and membership in G can be tested efficiently. Describe an algorithm which on input two permutations $\sigma, \pi \in G$ that are conjugate by an element in G, finds a $\tau \in G$ such that $\sigma = \tau^{-1}\pi\tau$. Can you analyze the complexity of your algorithm?*

Exercise 46 *Let p be prime. Do you think a (suitably formalized) version of the conjugacy decision problem in the finite group $\mathrm{GL}_n(\mathbb{F}_p)$ could be a suitable foundation for an ElGamal-type encryption scheme?*

Part III

Secret-Key Encryption

Chapter 7

Block ciphers

When large amounts of data have to be encrypted, public-key encryption is often considered as being too slow. High-speed solutions for encryption are commonly realized by means of cryptographic primitives that assume that the encrypting and decrypting party possess a common secret key. To establish such a common secret key across a public network, public-key encryption as discussed in Part II can be invoked to transport a symmetric key. Alternatively, a dedicated solution for key establishment can be used, which may offer additional advantages (cf. Chapter 9).

Arguably the most important building blocks within symmetric cryptography are so-called *block ciphers*.

DEFINITION 7.1 [Block cipher]   *Let k and n be positive integers. Then a* block cipher *with key length k and block size n is a family of permutations $\{\mathrm{Enc}_K : \mathbb{F}_2^n \longrightarrow \mathbb{F}_2^n \mid K \in \mathbb{F}_2^k\}$, i.e., each Enc_K can be regarded as an element of the symmetric group S_{2^n}.*

Differing from the situation for public-key encryption, for block ciphers it is common to have no provable reduction from a security claim to a (hopefully plausible) mathematical hardness assumption available. Offering strong provable security guarantees under plausible hardness assumptions and meeting the efficiency needs of high-speed or low-cost applications remains a design challenge. The so-called *ideal cipher model* captures the intuition behind a good block cipher: the ideal cipher is abstracted by the choice for each key $K \in \{0,1\}^k$ of the corresponding Enc_K uniformly at random in S_{2^n}. A typical choice for the block length is $n = 128$, making it completely impractical to write down a random permutation in S_{2^n}. For a well-designed block cipher, there should be no efficient way to make this distinction, however, as long as the secret key K remains completely unknown.

7.1   Advanced Encryption Standard

In 2000, the National Institute of Standards and Technology announced the outcome of a selection process for a new block cipher that can serve as national standard in the USA. This block cipher, originally referred to as *Rijndael*, is today widely known as AES. It is one of the most popular block ciphers, and on the following pages we describe its inner workings, following the pertinent standard document [NIS01].

Three different key lengths are possible—we can choose $k = 128$, $k = 192$, or $k = 256$. The main structure of the block cipher is not affected by this. Of course the change in key length has some effect on the details of the encryption and decryption process, but for instance the block length is always $n = 128$, independent of the key length.

7.1.1   Specifying the round function

To encrypt a plaintext $m \in \mathbb{F}_2^{128}$ we split m into 8-bit blocks and identify each of them with an element in the field $F_{256} = \mathbb{F}_2[x]/(x^8 + x^4 + x^3 + x + 1)$. Then we write these 8-bit blocks column-by-column into a square matrix. More specifically, we apply the following map ι to the plaintext m:

$$\iota : \quad \begin{array}{ccc} \mathbb{F}_2^{128} & \longrightarrow & F_{256}^{4 \times 4} \\ (m_0, \ldots, m_{127}) & \longmapsto & (s_{i,j})_{0 \leq i,j \leq 3} \end{array}, \tag{7.1}$$

with $s_{i,j} = \sum_{k=0}^{7} m_{8 \cdot (i+4j)+k} \cdot x^k \in F_{256}$.   The actual encryption will be performed by applying several rounds of invertible transformations to the state $S = (s_{i,j})_{0 \leq i,j \leq 3}$. The new state obtained after such a round depends on the input state and a round key. The individual round keys are derived from a secret k-bit key by means of a *key schedule* which we will discuss in a moment. But first, let us look at the four basic transformations used to form the round functions—none of them depends on the specific key length k, as round keys always end up as elements in $F_{256}^{4 \times 4}$. In fact, three of the four transformations are independent of the secret key.

SubBytes: This function substitutes each individual entry $s_{i,j}$ ($0 \leq i, j \leq 3$) of the state with a different value, derived from $s_{i,j}$ alone, i.e., these substitutions are key-independent and all of them can be performed in parallel. Specifically, SubBytes applies the map $\iota \circ \mathrm{aff} \circ \iota^{-1} \circ \mathrm{inv}$ to each entry $s_{i,j}$ of the state, with ι being the conversion map (7.1) and inv and aff being defined as follows:

$$\text{inv}: F_{256} \longrightarrow F_{256}$$
$$\alpha \longmapsto \begin{cases} \alpha^{-1} & , \text{ if } \alpha \neq 0 \\ 0 & , \text{ otherwise} \end{cases},$$

$$\text{aff}: \quad \mathbb{F}_2^8 \longrightarrow \mathbb{F}_2^8$$

$$\begin{pmatrix} b_0 \\ b_1 \\ b_2 \\ b_3 \\ b_4 \\ b_5 \\ b_6 \\ b_7 \end{pmatrix} \longmapsto \begin{pmatrix} 1 & 0 & 0 & 0 & 1 & 1 & 1 & 1 \\ 1 & 1 & 0 & 0 & 0 & 1 & 1 & 1 \\ 1 & 1 & 1 & 0 & 0 & 0 & 1 & 1 \\ 1 & 1 & 1 & 1 & 0 & 0 & 0 & 1 \\ 1 & 1 & 1 & 1 & 1 & 0 & 0 & 0 \\ 0 & 1 & 1 & 1 & 1 & 1 & 0 & 0 \\ 0 & 0 & 1 & 1 & 1 & 1 & 1 & 0 \\ 0 & 0 & 0 & 1 & 1 & 1 & 1 & 1 \end{pmatrix} \cdot \begin{pmatrix} b_0 \\ b_1 \\ b_2 \\ b_3 \\ b_4 \\ b_5 \\ b_6 \\ b_7 \end{pmatrix} + \begin{pmatrix} 1 \\ 1 \\ 0 \\ 0 \\ 0 \\ 1 \\ 1 \\ 0 \end{pmatrix}.$$

ShiftRows: This function is independent of the secret key as well and cyclically shifts each of the lower three rows by a row-dependent number of positions. More precisely, denoting by

$$\pi : \{0, 1, 2, 3\} \longrightarrow \{0, 1, 2, 3\}$$
$$j \longmapsto (j + 1) \bmod 4$$

a cyclic shift by one position to the left, ShiftRows is given by

$$(s_{i,j})_{0 \leq i,j \leq 3} \longmapsto (s_{i,\pi^i(j)})_{0 \leq i,j \leq 3}.$$

MixColumns: Just as the above two functions, MixColumns is independent of the secret key, and transforms the current state to a new one. As suggested by the name, this function operates column-wise—and all four columns can be dealt with in parallel. Specifically, the jth column is replaced according to the following F_{256}-linear map.

$$F_{256}^4 \longrightarrow F_{256}^4$$
$$\begin{pmatrix} s_{0,j} \\ s_{1,j} \\ s_{2,j} \\ s_{3,j} \end{pmatrix} \longmapsto \begin{pmatrix} x & x+1 & 1 & 1 \\ 1 & x & x+1 & 1 \\ 1 & 1 & x & x+1 \\ x+1 & 1 & 1 & x \end{pmatrix} \cdot \begin{pmatrix} s_{0,j} \\ s_{1,j} \\ s_{2,j} \\ s_{3,j} \end{pmatrix} \quad (0 \leq j \leq 3).$$

AddRoundKey: This is the only key-dependent transformation and takes next to the current state $(s_{i,j})_{0 \leq i,j \leq 3}$ a 128-bit key $(k_{i,j})_{0 \leq i,j \leq 3} \in F_{256}^{4 \times 4}$ as input. Then the state is updated according to

$$F_{256}^{4 \times 4} \longrightarrow F_{256}^{4 \times 4}$$
$$(s_{i,j})_{0 \leq i,j \leq 3} \longmapsto (s_{i,j} + k_{i,j})_{0 \leq i,j \leq 3}.$$

One complete *round* of AES takes a state $S \in F_{256}^{4 \times 4}$ and a 128-bit round key $\text{key} \in F_{256}^{4 \times 4}$ as input, and outputs

$$\text{AddRoundKey}(\text{MixColumns}(\text{ShiftRows}(\text{SubBytes}(S)))), \text{key}) \in F_{256}^{4 \times 4}. \quad (7.2)$$

To have a sufficient number of round keys available, AES relies on a key expansion which takes the secret key as input and derives 10, 12, or 14 round keys, depending on the length k of the secret key.

7.1.2 Key schedule

The number r of rounds used for an AES encryption is a function of the key length k, and we define $r = 6 + (k/32)$ in dependence on the key length $k \in \{128, 192, 256\}$. To enable a "whitening step" at the beginning of the encryption process, a total of $r + 1$ round keys are needed, each of which can be represented as a 4×4 matrix over F_{256}. To derive these round keys, we interpret the original k-bit secret key as a sequence $L_0, \ldots, L_{r-7}$ of eight vectors in F_{256}^4. This sequence is extended recursively according to the following rule:

$$L_i = L_{i-(r-6)} + f_i(L_{i-1}),$$

where

$$f_i : \; F_{256}^4 \longrightarrow \quad\quad\quad\quad\quad\quad\quad\quad F_{256}^4$$

$$\begin{pmatrix} a_0 \\ a_1 \\ a_2 \\ a_3 \end{pmatrix} \longmapsto \begin{cases} \begin{pmatrix} \text{SubBytes}(a_1) \\ \text{SubBytes}(a_2) \\ \text{SubBytes}(a_3) \\ \text{SubBytes}(a_0) \end{pmatrix} + \begin{pmatrix} x^{\frac{i}{r-6}-1} \\ 0 \\ 0 \\ 0 \end{pmatrix} & , \text{ if } i \bmod (r-6) = 0 \\[2em] \begin{pmatrix} \text{SubBytes}(a_0) \\ \text{SubBytes}(a_1) \\ \text{SubBytes}(a_2) \\ \text{SubBytes}(a_3) \end{pmatrix} & , \text{ if } k = 256 \text{ and } i \bmod (r-6) = 4 \\[2em] \begin{pmatrix} a_0 \\ a_1 \\ a_2 \\ a_3 \end{pmatrix} & , \text{ otherwise} \end{cases}$$

for $i = r - 6, \ldots, 4r - 1$. The resulting $4r$ column vectors $L_0, \ldots, L_{4r-1}$ are combined in groups of four in the natural way to obtain the required r round keys:

$$\text{key}_j = (L_{4j}|L_{4j+1}|L_{4j+2}|L_{4j+3}) \in F_{256}^{4 \times 4} \quad (j = 0, \ldots, r-1)$$

7.1.3 Encryption and decryption with AES

Knowing the inner workings of the individual AES rounds and how the round keys are derived, we are ready to describe the complete encryption procedure for a 128-bit plaintext block $m \in \mathbb{F}_2^{128}$. Figure 7.1 details the necessary steps, and decrypting the resulting ciphertext, using exactly the same secret key as for encryption, is straightforward, as each of the involved transformations is invertible (see Exercise 47).

ALGORITHM 7.1 AES encryption

Input: plaintext $m \in \mathbb{F}_2^{128}$
            secret key $K \in \{0,1\}^{128} \cup \{0,1\}^{129} \cup \{0,1\}^{256}$
Output: ciphertext $\mathrm{AES}_K(m) \in \mathbb{F}_2^{128}$

---

1. *Apply the key schedule to K to obtain session keys* $\mathrm{key}_0, \ldots, \mathrm{key}_{r-1}$.

2. *Initialize the internal state S to* $S \leftarrow \iota(m)$.

3. *Update the internal state to* $S \leftarrow \mathrm{AddRoundKey}(S, \mathrm{key}_0)$

4. *For $i = 1, \ldots, r-1$ update the internal state*
    $S \leftarrow \mathrm{AddRoundKey}(\mathrm{MixColumns}(\mathrm{ShiftRows}(\mathrm{SubBytes}(S))), \mathrm{key}_i)$.

5. *Update the internal state to* $S \leftarrow \mathrm{SubBytes}(S)$.

6. *Update the internal state to* $S \leftarrow \mathrm{ShiftRows}(S)$.

7. *Update the internal state to* $S \leftarrow \mathrm{AddRoundKey}(S, \mathrm{key}_r)$

8. *Return* $\mathrm{AES}_K(m) = \iota^{-1}(S)$ *as ciphertext.*

FIGURE 7.1: Encrypting with the Advanced Encryption Standard.

Group theory does not play a very prominent role in the analysis of block ciphers, and in this book we do not explore the details of the design of block ciphers. For readers who would like to get more acquainted with the design of block ciphers in general, *The Block Cipher Companion* by Knudsen and Robshaw [KR11] can serve as a starting point. For the specific case of AES, a book by the authors of the algorithm offers insight into the underlying design choices [DR02].

The round-based construction of AES is very typical for a block cipher, and having in mind the ideal cipher abstraction which we mentioned earlier, a question with a group-theoretic nature arises quite naturally. When we iterate over all possible session keys, which permutations on the plaintext space can we realize in this way? For AES specifically, we have a plaintext space of size 2^{128}, and in the ideal cipher model a session key results in a uniform random selection of all $2^{128}!$ possible permutations on this plaintext space. Even with the largest specified key length of 256 bits we certainly cannot hope to be able to realize *all* permutations on this plaintext space, as $2^{256} < 2^{128}!$.

It turns out that even with independently chosen round keys, the round functions of AES do not generate the complete symmetric group. More specifically, Wernsdorf showed that the set of round functions (7.2), when iterating over all possible round keys, generates a group that is isomorphic to the alternating group $A_{2^{128}}$ [Wer02]. We will not prove this statement completely here, but it is interesting to look in some more detail into working with a group whose elements are represented in such a non-standard way. For this, we give here one direction of Wernsdorf's proof.

THEOREM 7.1 [AES round functions are even]
Taken as permutations on $\mathbb{F}_2^{128}$, all of the following are even: SubBytes, ShiftRows, MixColumns, *and* AddRoundKey *(with an arbitrary but fixed round key). Consequently, the round functions of AES generate a subgroup of $A_{2^{128}}$.*

PROOF   Let us look at the four types of permutations in turn.

SubBytes: This permutation can be written as a product of 16 permutations, each of which captures the substitution on a single entry of the state and stabilizes the remaining 15 entries of the state. The permutation inv : $F_{256} \longrightarrow F_{256}$, which describes the substitution on one entry of the state, is an involution with two fixed points (0 and 1), so each of the 16 permutations comprising SubBytes can be written as a product of $127 \cdot 2^{120}$ transpositions. Being a product of $127 \cdot 2^{120} \cdot 2^4$ transpositions, SubBytes is even.

ShiftRows: This map is $\mathbb{F}_2$-linear and invertible. Thus we can represent ShiftRows as multiplication of an invertible matrix $M \in \mathbb{F}_2^{128 \times 128}$ with a binary vector of length 128 representing the current state. Using Gaussian elimination, we can decompose the matrix M into a product of matrices that differ from the 128×128 identity matrix only by a single off-diagonal entry. Each of these elementary matrices captures the addition of one row to another. Seen as permutation on $\mathbb{F}_2^{128}$, these elementary matrices have 2^{127} fixed points and split into a product of an even number (2^{126}) of transpositions. Hence ShiftRows must be even.

MixColumns: As this map is $\mathbb{F}_2$-linear as well, we can apply the same reasoning as for ShiftRows.

AddRoundKey: For the zero round key, we face the identity map, which clearly is even. For all other choices of the round key, addition of the round key translates into the composition of 2^{127} transpositions, which is obviously even.

Being a composition of even permutations, we recognize the round functions of AES as even. □

Similarly as for AES, the round functions of several other block ciphers have been shown to generate the alternating group, including the Data Encryption Standard (DES), one of the most influential block ciphers, which we will discuss in the next section. It deserves to be noted, however, that the size of the group generated by the round functions alone is not a good indicator of the strength of a block cipher. In an article from 1994, Murphy et al. gave an example of a block cipher whose round functions generate the full symmetric group on the plaintext space, but at the same time is rather weak [MPW94]. Notwithstanding this, understanding the group generated by the round functions of a block cipher is of value to rule out at least some major design flaws. The description of a block cipher may sometimes not lend itself to a convenient algebraic analysis, but for several major block ciphers it has been feasible to identify the group generated by their round functions.

7.2 Data Encryption Standard

From a modern perspective, the short effective key length of 56 bits immediately disqualifies the Data Encryption Standard (DES) for applications with high security requirements. As manifested in the key lengths of AES, currently a key length of at least 128 bits is expected, though 80 bit keys still enjoy some popularity when looking at low-cost platforms. Looking at the (by now withdrawn) standard publication specifying DES [NIS99], one notices that the "nominal" key length is actually 64 bit, but 8 of these have to satisfy a parity-check condition, which determines them uniquely once the remaining part of the key bits are fixed (see Section 7.2.3).

The plaintext and ciphertext space of DES is $\mathbb{F}_2^{64}$ and, similar to AES encryption is performed through iterating a round function which has a (64-bit) round key as input. A major difference is that in each round half of the input bits are only moved to a different position without changing their value. Only 32 bits are modified. This is due to the *Feistel structure* of DES.

7.2.1 General structure of DES: A Feistel cipher

The basic structure of one round of a *Feistel cipher* is depicted in Figure 7.2. The input from the previous round is split into two blocks which are swapped in the output of the round. Only one of the two input blocks is substituted through the addition of a value that depends on the round key and on the unmodified block.

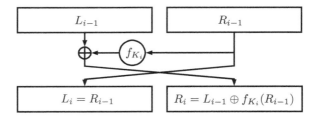

FIGURE 7.2:   One round of DES.

For DES specifically, the 64-bit input to the ith round is separated into its left and right halves $(L_{i-1}, R_{i-1}) \in \mathbb{F}_2^{32} \times \mathbb{F}_2^{32}$, and the output of this round is the concatenation of the 32-bit halves

$$(L_i, R_i) = (R_{i-1}, L_{i-1} \oplus f_{K_i}(R_{i-1})),$$

where f_{K_i} is a function that depends on a 48-bit round key K_i. While the specific choice of the function f_{K_i} affects the security of the resulting block cipher, it is important to note that the details of this function are insignificant for ensuring invertibility of the encryption process. In addition to several iterations of the round function, DES applies a key-independent initial permutation IP $: \mathbb{F}_2^{64} \longrightarrow \mathbb{F}_2^{64}$ to the 64-bit plaintext block at the beginning of the encryption. As this permutation is publicly known, for common attack models the initial permutation IP has no significance. In fact, a lightweight variant of DES that has been proposed in [LPPS07] dropped the initial and final permutation.

In the original DES, the initial permutation IP maps the 64-bit input block $(b_1, \ldots, b_{64}) \in \mathbb{F}_2^{64}$ to $(b_{\tau(1)}, \ldots, b_{\tau(64)}) \in \mathbb{F}_2^{64}$, where, in cycle notation, $\tau \in S_{64}$ is the following permutation of order 6 which leaves 22 and 43 fixed:

$$\begin{aligned}
\tau = \ &(1, 58, 55, 13, 28, 40)(2, 50, 53, 29, 32, 8)(3, 42, 51, 45, 27, 48) \\
&(4, 34, 49, 61, 31, 16)(5, 26, 56)(6, 18, 54, 21, 30, 24)(7, 10, 52, 37, 25, 64) \\
&(9, 60, 39)(11, 44, 35, 41, 59, 47)(12, 36, 33, 57, 63, 15)(14, 20, 38, 17, 62, 23) \\
&(19, 46).
\end{aligned}$$

The corresponding inverse permutation IP^{-1} is applied at the end of the encryption process. Postponing for a moment the details of the 16 round functions $f_{K_1}, \ldots, f_{K_{16}}$, a complete DES encryption is depicted in Figure 7.3.

The last round of the encryption algorithm is different from the previous 15 rounds in that the two processed halves are not swapped—the input to the final permutation is $R_{16} \| L_{16}$ rather than $L_{16} \| R_{16}$. This structure enables a very convenient implementation of the decryption algorithm: to decrypt a ciphertext with the secret key K, after deriving the round keys K_i, we relabel them so that K_{17-i} now becomes K_i for $i = 1, \ldots, 16$, i.e., we use the round

---

ALGORITHM 7.2 DES encryption

Input: *plaintext* $m \in \mathbb{F}_2^{64}$
   secret key $K \in \{0,1\}^{56}$
Output: ciphertext $\mathrm{DES}_K(m) \in \mathbb{F}_2^{64}$

---

1. *Apply the key schedule to K to obtain session keys $K_1, \ldots, K_{16}$.*

2. *Apply the initial permutation IP to m and split the result into its left half $L_0 \in \mathbb{F}_2^{32}$ and its right half $R_0 \in \mathbb{F}_2^{32}$.*

3. *for $i = 1, \ldots, 16$ compute*
   $(L_i, R_i) \leftarrow (R_{i-1}, L_{i-1} \oplus f_{K_i}(R_{i-1}))$

4. *Return $\mathrm{DES}_K(m) = \mathrm{IP}^{-1}(R_{16} \| L_{16})$ as ciphertext.*

FIGURE 7.3:   Encrypting with the Data Encryption Standard.

keys in reverse order. Then we proceed exactly as during encryption to recover the plaintext. Owing to the Feistel structure, correctness of this decryption process is independent of how exactly the key scheduling algorithm or the functions f_{K_i} look. Moreover, without knowing any details, an observation by Even and Goldreich [EG83] tells us that the round functions of DES realize an even permutation on the plaintext space $\mathbb{F}_2^{64}$:

THEOREM 7.2 [DES round functions are even]
Taken as permutations on $\mathbb{F}_2^{64}$, the round functions of DES are even.

PROOF   The round function $(L_i, R_i) \leftarrow (R_{i-1}, L_{i-1} \oplus f_{K_i}(R_{i-1}))$ can be decomposed into two permutations, both of which are even:

- A map add : $(L, R) \longmapsto (L \oplus f_{K_i}(R), R)$ modifying only the left half of its 64-bit input. For each $R \in \mathbb{F}_2^{32}$ with $f_{K_i}(R) = 0$, the map add leaves its input unchanged. For all R-values that are not mapped to 0^{32} by f_{K_i}, the values (L, R) and $(L \oplus f_{K_i}(R), R)$ are exchanged pairwise by add for any $L \in \mathbb{F}_2^{32}$. So with $\zeta = |f_{K_i}^{-1}(0^{32})|$ denoting the number of bitstrings that are mapped to $0 \in \mathbb{F}_2^{32}$ by f_{K_i}, we see that add can be written as a product of $2^{31} \cdot (2^{32} - \zeta)$ transpositions. Hence, add is even. (In the extreme case of f_{K_i} sending every input to 0^{32}, add is the identity permutation.)

- A map swap : $(L, R) \longmapsto (R, L)$ which exchanges the left half and the right half of its 64-bit input. This map fixes the 2^{32} elements with both halves being identical, and swaps the remaining $2^{64} - 2^{32}$ possible inputs pairwise. Therefore swap can be written as a product of $2^{31} \cdot (2^{64} - 1)$ transpositions, i. e., it is even.

$\square$

Invoking a slightly different (equivalent) representation of DES from Davio et al. [DDF$^+$84], Wernsdorf established the stronger result that the round functions of DES are not only even, but in fact generate a group that is isomorphic to the alternating group $A_{2^{64}}$ [Wer93].

7.2.2   Round function of DES

To fill in the details of the specific round function used in the Data Encryption Standard, we have to specify the function $f_{K_i} : \mathbb{F}_2^{32} \longrightarrow \mathbb{F}_2^{32}$ occurring in Figure 7.2. Given a 48-bit round key $K_i \in \mathbb{F}_2^{48}$, the corresponding function f_{K_i} in DES does the following.

1. The 32-bit input value $x = (x_1, \ldots, x_{32}) \in \mathbb{F}_2^{32}$ is expanded to the 48-bit value

$$
\begin{aligned}
X = (\ & x_{32}, x_1, \ x_2, \ x_3, \ x_4, \ x_5, \ x_4, \ x_5, \ x_6, \ x_7, \ x_8, \ x_9, \\
& x_8, \ x_9, \ x_{10}, x_{11}, x_{12}, x_{13}, x_{12}, x_{13}, x_{14}, x_{15}, x_{16}, x_{17}, \\
& x_{16}, x_{17}, x_{18}, x_{19}, x_{20}, x_{21}, x_{20}, x_{21}, x_{22}, x_{23}, x_{24}, x_{25}, \\
& x_{24}, x_{25}, x_{26}, x_{27}, x_{28}, x_{29}, x_{28}, x_{29}, x_{30}, x_{31}, x_{32}, x_1)
\end{aligned}
$$

2. Now the round key $K_i \in \mathbb{F}_2^{48}$ is added to the expanded input $X \in \mathbb{F}_2^{48}$. The resulting 48-bit value is split into eight 6-bit words $y^{(1)}, \ldots, y^{(6)} \in \mathbb{F}_2^{48}$, i. e., we have

$$ y^{(1)} || \ldots || y^{(6)} = X \oplus K_i. $$

3. Each value $y^{(i)}$ is substituted using the corresponding matrix $S^{(i)}$ in Figure 7.4. For $y^{(i)} = (y_0^{(i)}, y_1^{(i)}, y_2^{(i)}, y_3^{(i)}, y_4^{(i)}, y_5^{(i)})$, let $s(y^{(i)})$ be the entry of $S^{(i)}$ in

   - row no. $2 \cdot y_0^{(i)} + y_5^{(i)} \in \{0, \ldots, 3\}$ and
   - column no. $2^3 \cdot y_1^{(i)} + 2^2 \cdot y_2^{(i)} + 2 \cdot y_3^{(i)} + y_4^{(i)} \in \{0, \ldots, 15\}$.

4. Let $\mathrm{bin}(s(y^{(i)}))$ be the binary expansion of $s(y^{(i)})$ $(i = 1, \ldots, 8)$, e. g., $\mathrm{bin}(3) = (0, 0, 1, 1)$. Concatenating these binary expansions we obtain a 32-bit vector

$$ (z_1, \ldots, z_{32}) = \mathrm{bin}(s_1(y_1)) || \ldots || \mathrm{bin}(s_8(y_8)) \in \mathbb{F}_2^{32}. $$

   Now, $f_{K_i}(x) = (z_{\eta(1)}, \ldots, z_{\eta(32)}) \in \mathbb{F}_2^{32}$ where, in cycle notation, $\eta \in S_{32}$ is the following permutation:

$$
\begin{aligned}
\eta = \ & (1, 16, 10, 15, 31, 4, 21, 32, 25, 19, 24, 9) \\
& (2, 7, 28, 6, 12, 26, 13, 5, 29, 22, 27, 30, 11, 23, 3, 20, 14, 18, 8, 17)
\end{aligned}
$$

$$S^{(1)} = \begin{pmatrix} 14 & 4 & 13 & 1 & 2 & 15 & 11 & 8 & 3 & 10 & 6 & 12 & 5 & 9 & 0 & 7 \\ 0 & 15 & 7 & 4 & 14 & 2 & 13 & 1 & 10 & 6 & 12 & 11 & 9 & 5 & 3 & 8 \\ 4 & 1 & 14 & 8 & 13 & 6 & 2 & 11 & 15 & 12 & 9 & 7 & 3 & 10 & 5 & 0 \\ 15 & 12 & 8 & 2 & 4 & 9 & 1 & 7 & 5 & 11 & 3 & 14 & 10 & 0 & 6 & 13 \end{pmatrix}$$

$$S^{(2)} = \begin{pmatrix} 15 & 1 & 8 & 14 & 6 & 11 & 3 & 4 & 9 & 7 & 2 & 13 & 12 & 0 & 5 & 10 \\ 3 & 13 & 4 & 7 & 15 & 2 & 8 & 14 & 12 & 0 & 1 & 10 & 6 & 9 & 11 & 5 \\ 0 & 14 & 7 & 11 & 10 & 4 & 13 & 1 & 5 & 8 & 12 & 6 & 9 & 3 & 2 & 15 \\ 13 & 8 & 10 & 1 & 3 & 15 & 4 & 2 & 11 & 6 & 7 & 12 & 0 & 5 & 14 & 9 \end{pmatrix}$$

$$S^{(3)} = \begin{pmatrix} 10 & 0 & 9 & 14 & 6 & 3 & 15 & 5 & 1 & 13 & 12 & 7 & 11 & 4 & 2 & 8 \\ 13 & 7 & 0 & 9 & 3 & 4 & 6 & 10 & 2 & 8 & 5 & 14 & 12 & 11 & 15 & 1 \\ 13 & 6 & 4 & 9 & 8 & 15 & 3 & 0 & 11 & 1 & 2 & 12 & 5 & 10 & 14 & 7 \\ 1 & 10 & 13 & 0 & 6 & 9 & 8 & 7 & 4 & 15 & 14 & 3 & 11 & 5 & 2 & 12 \end{pmatrix}$$

$$S^{(4)} = \begin{pmatrix} 7 & 13 & 14 & 3 & 0 & 6 & 9 & 10 & 1 & 2 & 8 & 5 & 11 & 12 & 4 & 15 \\ 13 & 8 & 11 & 5 & 6 & 15 & 0 & 3 & 4 & 7 & 2 & 12 & 1 & 10 & 14 & 9 \\ 10 & 6 & 9 & 0 & 12 & 11 & 7 & 13 & 15 & 1 & 3 & 14 & 5 & 2 & 8 & 4 \\ 3 & 15 & 0 & 6 & 10 & 1 & 13 & 8 & 9 & 4 & 5 & 11 & 12 & 7 & 2 & 14 \end{pmatrix}$$

$$S^{(5)} = \begin{pmatrix} 2 & 12 & 4 & 1 & 7 & 10 & 11 & 6 & 8 & 5 & 3 & 15 & 13 & 0 & 14 & 9 \\ 14 & 11 & 2 & 12 & 4 & 7 & 13 & 1 & 5 & 0 & 15 & 10 & 3 & 9 & 8 & 6 \\ 4 & 2 & 1 & 11 & 10 & 13 & 7 & 8 & 15 & 9 & 12 & 5 & 6 & 3 & 0 & 14 \\ 11 & 8 & 12 & 7 & 1 & 14 & 2 & 13 & 6 & 15 & 0 & 9 & 10 & 4 & 5 & 3 \end{pmatrix}$$

$$S^{(6)} = \begin{pmatrix} 12 & 1 & 10 & 15 & 9 & 2 & 6 & 8 & 0 & 13 & 3 & 4 & 14 & 7 & 5 & 11 \\ 10 & 15 & 4 & 2 & 7 & 12 & 9 & 5 & 6 & 1 & 13 & 14 & 0 & 11 & 3 & 8 \\ 9 & 14 & 15 & 5 & 2 & 8 & 12 & 3 & 7 & 0 & 4 & 10 & 1 & 13 & 11 & 6 \\ 4 & 3 & 2 & 12 & 9 & 5 & 15 & 10 & 11 & 14 & 1 & 7 & 6 & 0 & 8 & 13 \end{pmatrix}$$

$$S^{(7)} = \begin{pmatrix} 4 & 11 & 2 & 14 & 15 & 0 & 8 & 13 & 3 & 12 & 9 & 7 & 5 & 10 & 6 & 1 \\ 13 & 0 & 11 & 7 & 4 & 9 & 1 & 10 & 14 & 3 & 5 & 12 & 2 & 15 & 8 & 6 \\ 1 & 4 & 11 & 13 & 12 & 3 & 7 & 14 & 10 & 15 & 6 & 8 & 0 & 5 & 9 & 2 \\ 6 & 11 & 13 & 8 & 1 & 4 & 10 & 7 & 9 & 5 & 0 & 15 & 14 & 2 & 3 & 12 \end{pmatrix}$$

$$S^{(8)} = \begin{pmatrix} 13 & 2 & 8 & 4 & 6 & 15 & 11 & 1 & 10 & 9 & 3 & 14 & 5 & 0 & 12 & 7 \\ 1 & 15 & 13 & 8 & 10 & 3 & 7 & 4 & 12 & 5 & 6 & 11 & 0 & 14 & 9 & 2 \\ 7 & 11 & 4 & 1 & 9 & 12 & 14 & 2 & 0 & 6 & 10 & 13 & 15 & 3 & 5 & 8 \\ 2 & 1 & 14 & 7 & 4 & 10 & 8 & 13 & 15 & 12 & 9 & 0 & 3 & 5 & 6 & 11 \end{pmatrix}$$

FIGURE 7.4: S-boxes of DES.

While the SubBytes step in AES is naturally specified in a compact manner as an algebraic operation in a finite field, the substitutions used in DES are commonly given directly as look-up tables in the form of the matrices $S^{(i)}$. These matrices are commonly referred to as *S-boxes*. The details of designing good S-boxes is a topic that we will not pursue in this book, but the reader who is interested in this topic may want to take a closer look at the lightweight version of DES proposed by Leander et al. in [LPPS07]—this DES variant replaces all eight S-boxes with a single new S-box, which is chosen to provide strong guarantees.

7.2.3 Key schedule

The nominal key length for DES is 64 bit, but every eighth bit of a secret key $(k_1, \ldots, k_{64}) \in \mathbb{F}_2^{64}$ cannot be chosen freely. Instead, those bits satisfy the parity check condition

$$k_{8i} = 1 + \sum_{j=1}^{7} k_{8i-j} \quad (i = 1, \ldots, 8).$$

Consequently, the effective key length of DES is only 56 bit, but in Exercise 50 we will see that a particular property of DES can be exploited to reduce an exhaustive key search to 2^{55} secret keys.

To form the 16 round keys from the given 64 bit key $(k_1, \ldots, k_{64})$, we start by forming two 28-bit subkeys

$$\begin{aligned}
C_0 = {} & (k_{57}, k_{49}, k_{41}, k_{33}, k_{25}, k_{17}, k_9, k_1, k_{58}, k_{50}, k_{42}, k_{34}, k_{26}, k_{18}, \\
& k_{10}, k_2, k_{59}, k_{51}, k_{43}, k_{35}, k_{27}, k_{19}, k_{11}, k_3, k_{60}, k_{52}, k_{44}, k_{36}) \text{ and} \\
D_0 = {} & (k_{63}, k_{55}, k_{47}, k_{39}, k_{31}, k_{23}, k_{15}, k_7, k_{62}, k_{54}, k_{46}, k_{38}, k_{30}, k_{22}, \\
& k_{14}, k_6, k_{61}, k_{53}, k_{45}, k_{37}, k_{29}, k_{21}, k_{13}, k_5, k_{28}, k_{20}, k_{12}, k_4).
\end{aligned}$$

In addition, we need a function : $\mathbb{F}_2^{28} \times \mathbb{F}_2^{28} \longrightarrow \mathbb{F}_2^{48}$ that selects a 48-round key on input of two 28-bit values. Specifically, with $C = (c_1, \ldots, c_{28})$ and $D = (d_1, \ldots, d_{28})$ we have

$$\begin{aligned}
p(C, D) = {} & (c_{14}, c_{17}, c_{11}, c_{24}, c_1, c_5, c_3, c_{28}, c_{15}, c_6, c_{21}, c_{10}, \\
& c_{23}, c_{19}, c_{12}, c_4, c_{26}, c_8, c_{16}, c_7, c_{27}, c_{20}, c_{13}, c_2, \\
& d_{13}, d_{24}, d_3, d_9, d_{19}, d_{27}, d_2, d_{12}, d_{23}, d_{17}, d_5, d_{20}, \\
& d_{16}, d_{21}, d_{11}, d_{28}, d_6, d_{25}, d_{18}, d_{14}, d_{22}, d_8, d_1, d_4).
\end{aligned}$$

Finally, we denote the cyclic left-rotation of $X = (x_1, \ldots, x_{28}) \in \mathbb{F}_2^{28}$ by $\text{rot}_k(X)$, e.g., $\text{rot}_2(X) = (x_3, x_4, \ldots, x_{28}, x_1, x_2)$. With this, the 16 round keys for DES are given as

$$
\begin{aligned}
K_1 &= p(\mathrm{rot}_1(C_0), \mathrm{rot}_1(D_0)) & K_2 &= p(\mathrm{rot}_2(C_0), \mathrm{rot}_2(D_0)) \\
K_3 &= p(\mathrm{rot}_4(C_0), \mathrm{rot}_4(D_0)) & K_4 &= p(\mathrm{rot}_6(C_0), \mathrm{rot}_6(D_0)) \\
K_5 &= p(\mathrm{rot}_8(C_0), \mathrm{rot}_8(D_0)) & K_6 &= p(\mathrm{rot}_{10}(C_0), \mathrm{rot}_{10}(D_0)) \\
K_7 &= p(\mathrm{rot}_{12}(C_0), \mathrm{rot}_{12}(D_0)) & K_8 &= p(\mathrm{rot}_{14}(C_0), \mathrm{rot}_{14}(D_0)) \\
K_9 &= p(\mathrm{rot}_{15}(C_0), \mathrm{rot}_{15}(D_0)) & K_{10} &= p(\mathrm{rot}_{17}(C_0), \mathrm{rot}_{17}(D_0)) \\
K_{11} &= p(\mathrm{rot}_{19}(C_0), \mathrm{rot}_{19}(D_0)) & K_{12} &= p(\mathrm{rot}_{21}(C_0), \mathrm{rot}_{21}(D_0)) \\
K_{13} &= p(\mathrm{rot}_{23}(C_0), \mathrm{rot}_{23}(D_0)) & K_{14} &= p(\mathrm{rot}_{25}(C_0), \mathrm{rot}_{25}(D_0)) \\
K_{15} &= p(\mathrm{rot}_{27}(C_0), \mathrm{rot}_{27}(D_0)) & K_{16} &= \qquad p(C_0, D_0))
\end{aligned}
$$

Owing to the short key length, a basic DES does not offer the security level one expects for many applications. Campbell and Wiener established that DES is not a group in the sense that encrypting twice

$$
m \longmapsto \mathrm{DES}_{K^{(2)}}(\mathrm{DES}_{K^{(1)}}(m))
$$

can in general not be replaced with a single DES encryption with some key $K^{(3)}$ [CW93]. At the time of writing this book, a particular type of iterated application of the basic DES still enjoys quite some popularity—*Triple DES* uses three DES keys $(K^{(1)}, K^{(2)}, K^{(3)})$. To encrypt a 64-bit plaintext m, the ciphertext

$$
c = \mathrm{DES}_{K^{(3)}}(\mathrm{DES}_{K^{(2)}}^{-1}(\mathrm{DES}_{K^{(1)}}(m)))
$$

is computed. So the plaintext is first encrypted under the key $K^{(1)}$, this intermediate ciphertext is then decrypted under key $K^{(2)}$, and the result of this operation is finally encrypted under key $K^{(3)}$. Running DES in decryption mode with $K^{(2)}$ ensures a convenient interoperability with "Single DES" encryption: when all three keys are chosen to be the same, i.e., $K^{(1)} = K^{(2)} = K^{(3)}$, then encrypting with Triple DES is equivalent to encrypting with DES using the key $K^{(1)}$. Another keying option for Triple DES is to choose $K^{(1)}$ and $K^{(2)}$ independently, and to impose $K^{(3)} = K^{(1)}$.

7.3 Permutation Group Mappings

In the same decade when DES became a national standard, S. Magliveras considered a proposal for building a block cipher by means of logarithmic signatures for a finite group G. We took a look at logarithmic signatures in Section 4.6 already, and the core idea of the block cipher PGM is to use the map

$$
\begin{aligned}
\pi_{\alpha,\beta} : \mathbb{Z}_{|G|} &\longrightarrow \mathbb{Z}_{|G|} \\
x &\longmapsto \mu_{(r_1', \dots, r_{s'}')}^{-1} \circ f_\beta^{-1} \circ f_\alpha \circ \mu_{(r_1, \dots, r_s)}^{-1}(x) \cdot
\end{aligned} \tag{7.3}
$$

from Equation (4.5) as encryption. Differing from the asymmetric setting, there is no need to publish information about the pair of logarithmic signatures (α, β)—possibly apart from the *size* of the underlying group G. The latter characterizes the plaintext space completely, and there is no need to reveal details of G.

To be able to actually encrypt a plaintext in $\mathbb{Z}_{|G|}$, the logarithmic signature β is chosen in a way that ensures the availability of an efficient algorithm for factoring along β. Moreover, to facilitate decryption, the logarithmic signature α is ensured to afford an efficient method for factoring as well.

REMARK 7.1 Providing an efficient decryption operation may seem like a trivial necessity for a block cipher. It deserves to be noted, however, that not all modes of operation (see Section 7.4) make use of the decryption operation. So even in the absence of an efficient decryption algorithm, a block cipher may be of help for encrypting and decrypting plaintexts. □

When dealing with permutation groups $G \le S_n$ with n not being too large, a construction from [MM92] ensures the availability of a logarithmic signature where factoring group elements is not costly. Specifically, consider the stabilizer chain

$$G = G_0 \ge G_1 \ge \cdots \ge G_{n-1} = \{\mathrm{id}\},$$

where G_i is the subgroup of G consisting of all $\pi \in G_i$ with—at least—the i fixed points $\pi(1) = 1, \ldots, \pi(i) = i$. Now, for each $i \in \{1, \ldots, n\}$, we define $\alpha_i = [\alpha_{i0}, \ldots, \alpha_{ir_i-1}] \in G_{i-1}^{r_i}$ as a non-empty sequence of minimal length such that $\{\alpha_{i0}(i), \ldots, \alpha_{ir_i-1}(i)\}$ equals the orbit $\{\pi(i) : \pi \in G_{i-1}\}$ of i under the natural permutation action of G_{i-1}. Then G_{i-1} is the disjoint union of the cosets $\alpha_{i1} \cdot G_i, \ldots, \alpha_{ir_i-1} \cdot G_i$, and deciding for $\phi \in G_{i-1}$ to which coset it belongs reduces to identifying which of the (at most n candidate) indices j_i satisfies $\phi(i) = \alpha_{ij_i}(i)$.

If $\phi \in \alpha_{ij_i} \cdot G_i$, then the composition $\alpha_{ij_i}^{-1} \cdot \phi$ is contained in G_i. So iterating through $i = 1, \ldots, n$, for a given group element ϕ we can derive an expression

$$\alpha_{nj_n}^{-1} \cdot \cdots \cdot \alpha_{1j_1}^{-1} \cdot \phi \in G_{n-1} = \{\mathrm{id}\},$$

or equivalently $\phi = \alpha_{1j_1} \cdot \cdots \cdot \alpha_{nj_n}$.

Certain choices of α and β certainly have to be avoided to obtain a secure PGM instance. For instance, if we choose these two logarithmic signatures to be equal, then encryption becomes the identity map. According to Magliveras and Memon in [MM92], the logarithmic signatures α and β are in practice chosen to be of different *type*. Here the *type* of a logarithmic signature $\alpha = [\alpha_1, \ldots, \alpha_s]$ with $\alpha_i \in G^{r_i}$ is the vector $(r_1, \ldots, r_s)$. It is worth noting that almost 50 years after its invention no structural weakness in PGM has been published, and several interesting algebraic properties of this block cipher have been established [ibid.]. Being of comparable age as the Data Encryption Standard, it seems fair to say that this block cipher offers one of the most long-standing cryptanalyic challenges among group-theoretic designs in cryptography.

7.4 Modes of operation

By design, a block cipher encrypts only a (short) message block of a fixed length n, and it needs to be clarified how a larger amount of data, or a plaintext *stream*, is to be handled. This is addressed by the *mode of operation* of a block cipher. Conceptually it is important to note that a mode of operation treats the block cipher as a black box, i. e., the inner workings are not significant. In the subsequent sections, we will take a brief look at five common ways block ciphers are used when dealing with larger amounts of data. More details on these modes can be found in [Dwo01], and it should be noted that our discussion of modes of operation is not exhaustive. For instance, our discussion does not cover the *Galois/counter (GCM) mode*, which is specified in [Dwo07].

7.4.1 Electronic codebook (ECB) mode

Perhaps the most obvious way to deal with larger amounts of data is to divide the data into consecutive pieces which are then encrypted individually, either processing one by one or, if multiple instances of the block cipher are available, several independently in parallel (see Figure 7.5).

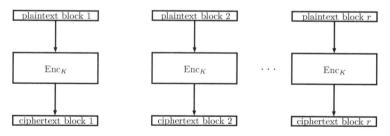

FIGURE 7.5: Encryption in electronic codebook (ECB) mode.

Assuming that the block cipher operates on n-bit inputs and the plaintext consists of r blocks $m_1, \ldots, m_r \in \mathbb{F}_2^n$ of identical size n, the *Electronic codebook*

(ECB) mode produces the ciphertext sequence $c_1, \ldots, c_r$ where

$$c_i = \mathrm{Enc}_K(m_i) \quad (i = 1, \ldots, r),$$

with some secret key K. Decryption is done in the obvious way, invoking the decryption operation of the block cipher:

$$m_i = \mathrm{Enc}_K^{-1}(c_i) \quad (i = 1, \ldots, r).$$

It is important to realize that encrypting identical plaintext blocks m_i in ECB mode results in identical ciphertext blocks c_i, which in some application contexts may compromise confidentiality (see Exercise 51). Figure 7.5 illustrates the encryption process for multiple plaintext blocks when using the ECB mode.

7.4.2   Cipher block chaining (CBC) mode

Differing from the ECB mode we just discussed, in the *cipher block chaining (CBC) mode*, the ciphertext corresponding to an individual plaintext block m_i depends on the result of the encryption of the previously encrypted plaintext block m_{i-1}. To handle the first message block, an n-bit initialization vector $c_{-1} \in \mathbb{F}_2^n$ is used. More specifically, the CBC mode can be applied to a plaintext consisting of blocks $m_1, \ldots, m_r \in \mathbb{F}_2^n$ of identical size n and results in a ciphertext sequence $c_1, \ldots, c_r$ of the same length, where

$$c_i = \mathrm{Enc}_K(m_i + c_{i-1}) \quad (i = 1, \ldots, r).$$

Here $m_i + c_{i-1}$ denotes addition in $\mathbb{F}_2^n$. The initialization vector must not be predictable, but it is not necessary to keep it secret. Invoking the decryption operation of the block cipher and having the correct initialization vector available, the plaintext can be recovered in the obvious manner:

$$m_i = c_{i-1} + \mathrm{Enc}_K^{-1}(c_i) \quad (i = 1, \ldots, r)$$

Figure 7.6 illustrates the processing of multiple plaintext blocks when encrypting in CBC mode.

7.4.3   Cipher feedback (CFB) mode

Similar to the CBC mode, for using the *cipher feedback (CFB) mode*, an unpredictable initialization vector $e_1 \in \mathbb{F}_2^n$ has to be provided when working with a block cipher that operates on n-bit blocks, and it is not required to keep e_1 secret. When using the CFB mode, a parameter $s \in \{1, \ldots, n\}$ needs to be specified which determines the size of the plaintext blocks $m_1, \ldots, m_r \in \mathbb{F}_2^s$ that are to be encrypted. The term *s-bit CFB mode* can be used to make the choice of s explicit. The encryption process is illustrated in Figure 7.7 and

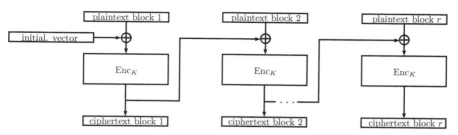

FIGURE 7.6: Encryption in cipher block chaining (CBC) mode.

can be described as follows:

$$c_i = m_i + [\text{Enc}_K(e_i)]_{1\to s} \quad , \text{ where}$$
$$e_{i+1} = [e_i]_{s+1\to n}||c_i \quad (i = 1, \ldots, r).$$

At this, $||$ denotes concatenation and $[(\alpha_1, \ldots, \alpha_n)]_{i\to j} = (\alpha_i, \ldots, \alpha_j)$ projection on the indicated components.

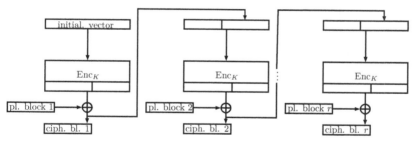

FIGURE 7.7: Encryption in (s-bit) cipher feedback (CFB) mode.

Differing from ECB mode and CBC mode, for recovering the plaintext from the ciphertext, the decryption operation of the underlying block cipher is not needed. Starting with the initialization vector e_1, we can compute the "masking values" e_i in the same way as during encryption:

$$m_i = c_i + [\text{Enc}_K(e_i)]_{1\to s} \quad , \text{ where}$$
$$e_{i+1} = [e_i]_{s+1\to n}||c_i \quad (i = 1, \ldots, r).$$

7.4.4 Output feedback (OFB) mode

The initialization vector e_0 used in the *output feedback (OFB) mode* must be used only once (a nonce). This initialization vector serves as initial seed to produce a (pseudo-random) sequence that is added to the plaintext blocks, one by one. This type of encryption is commonly used in *stream ciphers*, which are designed to produce pseudo-random sequences at high speed. On input a sequence of plaintext blocks $m_1, \ldots, m_r$, the OFB mode outputs a sequence of ciphertext blocks $c_1, \ldots, c_r$ of the same length, but it is not necessary that the overall length of the plaintext is a multiple of the input size n of the underlying block cipher. Suppose that $m_1, \ldots, m_{r-1} \in \mathbb{F}_2^n$ are n-bit blocks and $m_r \in \mathbb{F}_2^{n-\delta}$ with $\delta \in \{0, \ldots, n-1\}$. Then the resulting ciphertext is computed as follows—keeping with the projection notation from Section 7.4.3:

$$c_i = m_i + e_i \quad (i = 1, \ldots, r-1) \quad \text{and}$$
$$c_r = m_r + [e_r]_{1 \to n-\delta} \quad , \text{where}$$
$$e_i = \text{Enc}_K(e_{i-1}) \quad (i = 1, \ldots, r)$$

Figure 7.8 illustrates the encryption of multiple plaintext blocks in this mode, assuming that all plaintext blocks blocks have a length of n bits.

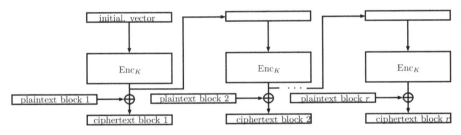

FIGURE 7.8:   Output feedback (OFB) mode.

Decryption is performed in the obvious way, making use of the initialization vector e_0—and without invoking the decyption operation of the underlying block cipher:

$$m_i = c_i + e_i \quad (i = 1, \ldots, r-1) \quad \text{and}$$
$$m_r = c_r + [e_r]_{1 \to n-\delta} \quad , \text{where}$$
$$e_i = \text{Enc}_K(e_{i-1}) \quad (i = 1, \ldots, r)$$

7.4.5   Counter (CTR) mode

Just as in OFB mode, in *counter (CTR) mode* we permit plaintexts consisting of $r-1$ consecutive blocks $m_1, \ldots, m_{r-1} \in \mathbb{F}_2^n$ of "full length" followed by a last block $m_r \in \mathbb{F}_2^{n-\delta}$ ($\delta \in \{0, \ldots, n-1\}$) which is potentially shorter. To employ the CTR mode, r different *counters* $e_1, \ldots, e_r \in \mathbb{F}_2^n$ are needed—they have to be pairwise different across all plaintext blocks for which the same secret key K is used. Similarly as in OFB mode, the block cipher is never applied to the individual plaintext blocks directly:

$$c_i = m_i + \text{Enc}_K(e_i) \quad (i = 1, \ldots, r-1) \quad \text{and}$$
$$c_r = m_r + [e_r]_{1 \to n-\delta}$$

For the case that all plaintext blocks have length n and the e_i-values are derived from a (random) initial value with a simple counter (mod 2^n), the encryption process is illustrated in Figure 7.9.

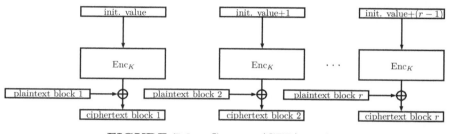

FIGURE 7.9: Counter (CTR) mode.

The decryption process is straightforward. Knowing the e_i-values, the recipient has the option to precompute the masking values and to parallelize decyption:

$$m_i = c_i + \text{Enc}_K(e_i) \quad (i = 1, \ldots, r-1) \quad \text{and}$$
$$m_r = c_r + [e_r]_{1 \to n-\delta}$$

7.5 Summary and further reading

With the Advanced Encryption Standard and its predecessor, the Data Encryption Standard, we have seen two prominent block ciphers. Both of them have a round-oriented structure, with each round depending on a round key which is derived from an initial key through a key schedule. As permutations on the plaintext space, the round functions of AES and DES are even. With Permutation Group Mappings, we have seen an, as of now unbroken, suggestion for deriving a block cipher by means of a group-theoretic construction. Finally, we have seen five different modes of operation for block ciphers, and in several of these modes neither encryption nor decryption require the decryption operation of the underlying block cipher.

As early as in 1975, Coppersmith and Grossman suggested to look at the size of the group generated by the round functions as a measure for the strength of a block cipher [CG75]. Paterson demonstrated that next to the size of this group, more fine-grained properties can be of interest—he demonstrates that a weakness can arise if the group generated by the round functions acts *imprimitively* on the message space [Pat99].

Block ciphers derived from factorizations in finite groups have been explored by several authors since the original PGM proposal, including implementation aspects. For instance Horváth et al. [HMvT94] report on a hardware implementation of permutation multiplication as occurring in PGM and Čanda et al. report on software experiments [ČvTMH01]. It seems fair to say, however, that the main interest in these constructions is so far within academia. At the time of writing this book, AES can be seen as the most popular block cipher, but there is interest in alternative designs, especially if implementations fit on platforms with very limited hardware resources. One example of such a lightweight block cipher is PRESENT [BKL+07].

Efficient and secure modes of operation for block ciphers play an important role in securing data, and the interest in identifying such modes goes beyond academia. Our discussion covered only the most common modes and we focused on securing the confidentiality of the transmitted data. The Galois/Counter Mode (GCM) [Dwo07] briefly mentioned in the text, on the other hand, takes the question of authenticity explicitly into account as well.

7.6   Exercises

Exercise 47
Specify transformations InvSubBytes, InvShiftRows, InvMixColumns, *and* InvAddRoundKey *such that all of the following maps are the identity on their respective domain of definition:*

> InvSubBytes ∘ SubBytes
> InvShiftRows ∘ ShiftRows
> InvMixColumns ∘ MixColumns
> InvAddRoundKey ∘ AddRoundKey *(for any fixed round key)*

Then give a decryption procedure InvAES *with* $\text{InvAES}_K(\text{AES}_K(m)) = m$ *for all* $m \in \mathbb{F}_2^{128}$ *and all secret keys* $K \in \{0,1\}^{128} \cup \{0,1\}^{192} \cup \{0,1\}^{256}$.

Exercise 48 *Is the permutation* inv *on* F_{256}, *which* SubBytes *applies to each entry of the current state of AES, an even or an odd permutation?*

Exercise 49 *Prove that the group generated by the AES round functions is transitive on* $F_{256}^{4\times4}$, *i.e., for each pair of states* (S, S') *we can find a group element that transforms* S *into* S'.

Exercise 50 *Suppose we are given a 64-bit plaintext m and the corresponding ciphertext $c = \mathrm{DES}_K(m)$ with an unknown secret key K. Show that from such a pair (m, c) we obtain immediately a plaintext-ciphertext pair (m', c') for a second secret key $K' \neq K$. Can this be exploited in an exhaustive key search?*

Exercise 51 *Suppose that a 1024×768 picture is stored in a file format that encodes the color value of each pixel into a 32-bit word, storing the pixels row-wise. So four neighboring pixels fit into one 128-bit block. Explain why using the ECB mode for encrypting this type of data can be problematic.*

Exercise 52 *When analyzing a construction like Triple DES, one naturally encounters the concept of a* pure *set of permutations. Let $n \geq 2$ and $E \subseteq S_n$ be a non-empty set of permutations [KJRS86, BvL13]. We call E* pure *if for all $e, f, g \in E$ there is $h \in E$ such that $h = g \circ f^{-1} \circ e$. For a construction like Triple DES, if we choose E as the set of permutations on the plaintext space induced by the underlying block cipher when iterating over all possible keys, then E being* pure *is not a desirable property.*

(a) *Show that a pure subset $E \subseteq S_n$ is a group if and only if E contains the identity.*

(b) *Show that every coset of the alternating group $A_n \trianglelefteq S_n$ is pure.*

Exercise 53 *Determine two logarithmic signatures α, β for the symmetric group S_4 that are not of the same type, and encrypt $0 \in \mathbb{Z}/24\mathbb{Z}$ using PGM with (α, β) as secret key.*

Exercise 54 *Let $m_1, \ldots, m_r \in \mathbb{F}_2^{128}$ be a sequence of consecutive plaintext blocks. We apply AES in CFB mode with parameter $s = 128$ to derive a sequence of ciphertext blocks $c_1, \ldots, c_r$. Now, suppose one bit in the first ciphertext block c_1 is flipped, because of an (undetected) error. Which ciphertext blocks are still decrypted correctly?*

Chapter 8

Cryptographic hash functions and message authentication codes

In the discussion of signature schemes in Chapter 10, cryptographic *hash functions* will play an important role, but they are brought to use in many other cryptographic applications as well. Unlike encryption schemes, hash functions do not involve any type of secret key, but nonetheless they form an important building block in many cryptographic constructions. While proposals for hash functions have been made that try to exploit the algebraic structure inherent to a (particular representation of a) finite group, at this point such designs do not play a prominent role outside academia. Also, similarly as for block ciphers, the role of proofs is not given as much weight as in the design of asymmetric encryption schemes. Because of hash functions being potentially applied to very large datasets, computational efficiency traditionally gets a lot of attention in the design process.

8.1 Cryptographic hash functions

At its core, a cryptographic hash function is a function h with a finite codomain and a domain that has a larger cardinality than the codomain. Such a function can by definition not be injective, yet it is one of the four essential design requirements that "for practical purposes" we can think of h as being one-to-one:

Efficiency. Evaluating h must be efficient, i.e., it should be possible to process large inputs—preferably even on platforms with very limited computational resources.

Preimage resistance. Given a value h_0 in the image of h, obtained by applying h to a "random" input, it must not be feasible to recover a preimage of h_0.

Second-preimage resistance. Given a value h_0 in the image of h along with a preimage m_0, it must not be feasible to find another preimage $m_1 \neq m_0$ satisfying $h(m_1) = h_0$.

Collision resistance. It must not be feasible to find any two different inputs $m_0 \neq m_1$ such that $h(m_0) = h(m_1)$. So the (short) function value $h(m)$ can serve as unique "fingerprint" for a (possibly long) message m.

In this introductory text, an informal description of these requirements will suffice, but to establish precise provable guarantees, more technical work is needed (cf. the definition of collision-resistant hash functions in Section 5.1). For the reader who is interested in formalizing hash function requirements and understanding the challenges in finding appropriate technical definitions, work by Rogaway and Shrimpton [RS04], Rogaway [Rog06], and Andreeva and Stam [AS11] can serve as starting points.

By design, hash functions do not involve any form of secret key, and their specification must be public. In this sense, the reader may rightfully ask why we discuss them in a chapter on symmetric encryption. As will be seen in Section 8.4, hash functions can be used to derive message authentication codes, and these can in turn be used to augment symmetric encryption schemes with guarantees beyond confidentiality of the plaintext. It is fair to say that the usual "real-life approximation" of a random oracle (cf. Section 4.3) is a (by definition deterministic) hash function. Before looking at some specific examples of suggested designs, it is worth noting that the collision resistance requirement imposes a lower bound on the output size that should be considered. The following result, due to Sayrafiezadeh [Say94], explains why currently used hash functions have an output length of at least 160, or even 256, bits.

THEOREM 8.1 [Birthday phenomenon]
If we sample, independently and uniform random, k times from a finite set of size n with $1 < k \leq n$, then the probability of drawing at least twice the same element is greater than

$$1 - \left(1 - \frac{k}{2n}\right)^{k-1} . \tag{8.1}$$

PROOF   There are $n \cdot (n-1) \cdot \cdots \cdot (n-k+1)$ ways to draw k pairwise different values from the set. Therefore, the exact probability of the event we look for is

$$1 - \frac{n \cdot (n-1) \cdot \cdots \cdot (n-k+1)}{n^k},$$

and it is sufficient to establish a suitable upper bound for

$$\frac{(n-1) \cdot \cdots \cdot (n-k+1)}{n^{k-1}} = \prod_{i=1}^{k-1} \left(1 - \frac{i}{n}\right) . \tag{8.2}$$

The inequality between the arithmetic and the geometric mean ensures

$$\sqrt[k-1]{\prod_{i=1}^{k-1}\left(1-\frac{i}{n}\right)} < \underbrace{\frac{1}{k-1}\cdot\sum_{i=1}^{k-1}\left(1-\frac{i}{n}\right)}_{=1-\frac{k}{2n}}.$$

Now, raising both sides of this inequality to the power $k-1$ we obtain the required upper bound on the term (8.2). □

Let us apply this theorem to an idealized hash function—a random oracle H—that maps inputs onto a set of size n, and assume we can query H with about $k = c\cdot\sqrt{n}$ different values for some positive constant c. With a typical value of n being 2^{160} or larger, it seems plausible to evaluate the lower bound (8.1) asymptotically for large n:

$$\lim_{n\to\infty} 1 - \left(1-\frac{c\cdot\sqrt{n}}{2n}\right)^{c\sqrt{n}-1} = 1 - \lim_{n\to\infty}\left(1-\frac{c}{2\sqrt{n}}\right)^{c\sqrt{n}-1} = 1 - \frac{1}{e^{c^2/2}}.$$

So already for $k > \sqrt{2\ln 2}\cdot\sqrt{n}$ we expect the chance for encountering a collision to be at least 50%.

For preimage resistance, the situation is less advantageous for an adversary. Consider a random oracle, mapping onto a finite set of size n. Then the probability for any particular input to map on a fixed, given, output value is $1/n$. The probability of finding a preimage after querying k pairwise different values is

$$1 - \left(\frac{n-1}{n}\right)^{k} = 1 - (1-1/n)^{k}. \tag{8.3}$$

In particular, if we allow again $k = c\cdot n^{\alpha}$ queries with a positive constant c and $0 < \alpha < 1$, this probability converges to 0 when n approaches infinity (Exercise 55).

8.2 Deriving a hash function from a block cipher

In real life implementations hash functions are indeed not random oracles, and so the question arises on what a hash function actually looks like. There are dedicated constructions; the SHA-2 and SHA-3 families in the Secure Hash Standard [NIS12] and the SHA-3 Standard [NIS14] offering prominent examples, but it is also possible to derive hash functions from a block cipher. Especially if a target platform offers hardware support for a block cipher, this can be of interest—for AES such support is rather widespread. A basic challenge that has to be overcome is the fixed block width and key size of a block cipher. Throughout we will assume that the key length is equal to the block length n of the block cipher, and we will assume that n is large enough for collision resistance not being thwarted by the birthday phenomenon.

As a hash function must be able to handle inputs of varying (arbitrary) length, whereas the block cipher is restricted to inputs of fixed length, in a first step the message that is to be hashed is *padded*. After padding, the message length is guaranteed to be a multiple of n. To realize this without making a collision search trivial, a common technique is to append bits to the original message in a deterministic manner, followed by appending a block of fixed length that reflects the length of the original message. This technique is commonly known as *Merkle–Damgård strengthening*.

Example 8.1
Here is a popular variant, where $\gamma \geq 0$ is a fixed parameter, e. g., $\gamma = 128$.

- Append a single 1-bit to m: $m||1$.

- Append a minimal non-negative number j of 0-bits to $m||1$, so that the length of $m||1||0^{j+\gamma}$ is divisible by n.

- Output $m||1||0^j||c$ where c is the unique binary γ-bit representation of the length of the original input m in binary.

Obviously such a padding restricts the input *length* of m to at most $2^\gamma - 1$, but already a value of $\gamma = 128$ suffices to satisfy rather extreme requirements for input sizes. □

Next, we need a *compression function* $f : \{0,1\}^n \times \{0,1\}^n \longrightarrow \{0,1\}^n$. As shown in Figure 8.1, a compression function along with a fixed (and public) initialization value h can then be used to process an input block by block, finally obtaining a hash value of length n. This design principle is named after Merkle and Damgård, and we are interested in deriving the compression function f from a block cipher. Preneel et al. [PGV94] and Black et al. [BRS02] explored 64 natural candidates to derive the compression function f from a block cipher. One of the twelve constructions that did not succumb to an attack is due to Matyas et al. [MMO85]:

$$f(h, m) = E_h(m) \oplus m \oplus h$$

where $E_K(\cdot)$ denotes encryption with the block cipher under key K, and $\oplus$ addition in $\mathbb{F}_2^n$. Another popular construction is commonly named after Davies and Meyer, despite Davies himself apparently not claiming to be a proposer of this scheme [PGV94]:

$$f(h, m) = E_m(h) \oplus h \tag{8.4}$$

ALGORITHM 8.1 Hashing with a compression function

Input: *bitstring* $(m_1, \ldots, m_r) \in \mathbb{F}_2^{n \cdot r}$ *with* $r \geq 0$
initialization vector $h \in \mathbb{F}_2^n$
Output: *hash value* $h \in \mathbb{F}_2^n$

---

1. *for* $i = 1, \ldots, r$
   replace $h \leftarrow f(h, m_i)$

2. *output* h

FIGURE 8.1: Applying the Merkle–Damgård design with a compression function $f : \mathbb{F}_2^n \times \mathbb{F}_2^n \longrightarrow \mathbb{F}_2^n$, assuming input lengths to be a multiple of n.

It is worth noting that for the Davies–Meyer construction one can efficiently construct fixed points in the sense that for each $h \in \mathbb{F}_2^n$ one can efficiently find a message m_0 that satisfies $f(h, m_0) = h$ (see Exercise 57).

8.3 Cayley hash functions

Let G be a finite group and $S = \{g_0, \ldots, g_{t-1}\}$ a set of generators for G. Then we can define a surjective map

$$h : \{0, \ldots, t-1\}^* \longrightarrow G$$
$$(m_1, \ldots, m_r) \longmapsto g_{m_1} \cdot \cdots \cdot g_{m_r}$$

In particular, for a generating set with two elements, we obtain a map that assigns to each bitstring an element of the group G. Intuitively, applying h describes a walk on a Cayley graph of G.

DEFINITION 8.1 [Cayley graph] *Let G be a finite group and S a set of generators for G. Then a (directed) graph with vertex set G and edge set $\{(g, g \cdot s) \in G \times G : s \in S\}$ is called a Cayley graph for G.*

The mapping h is a basic version of a *Cayley hash* [PLQ07]. With the interpretation of applying h as a walk in a Cayley graph, one can try to use graph-theoretic insights to obtain information on the security of h. For instance, let γ be the length of the shortest cycle—the *girth*—in the undirected version of the Cayley graph. Then the lengths of two colliding inputs $m \neq m'$, i.e., inputs with $h(m) = h(m')$, must satisfy $|m_1| + |m_2| \geq \gamma$.

Cayley hash functions have a remarkable (homomorphic) property in regard to concatenation—for any inputs m_1, m_2 we have $h(m_1 || m_2) = h(m_1) \cdot h(m_2)$. Moreover, parallelizing the evaluation of such a hash function is straightforward, as multiplication in the group G is associative. Differing from the Merkle–Damgård design with a compression function shown in Figure 8.1, making use of multiple available processors to speed up the hashing process is quite obvious. For bringing this appealing idea to use, one has to fix specific group generators and a concrete representation.

One of the most prominent constructions in the family of Cayley hash functions is due to J.-P. Tillich and G. Zémor and dates back to the end of the 20th century [TZ94]. The platform group used in this proposal is the special linear group $\mathrm{SL}_2(\mathbb{F}_{2^m})$ of 2×2 matrices with determinant 1 over a binary field $\mathbb{F}_2^m$. The size of this group computes to $2^{3m} - 2^m$, and the proposed values for m are (prime numbers) in the range $130 \leq m \leq 170$. A motivation for choosing m to be prime is to thwart attacks that make use of intermediate fields between $\mathbb{F}_2$ and $\mathbb{F}_{2^m}$. Using a polynomial basis representation, one can represent elements in $\mathrm{SL}_2(\mathbb{F}_{2^m})$ quite compactly as a sequence of four $\mathbb{F}_{2^m}$-elements. Exploiting that the determinant of each involved matrix is 1, one can try to reduce the size of hash values even further.

Let us write $\mathbb{F}_{2^m} = \mathbb{F}_2[x]/(f(x))$ with an irreducible polynomial $f(x) \in \mathbb{F}_2[x]$ of degree m, and let $\alpha \in \mathbb{F}_{2^m}$ be a root of $f(x)$. With this notation, the generators suggested by Tillich and Zémor are

$$g_0 = \begin{pmatrix} \alpha & 1 \\ 1 & 0 \end{pmatrix} \text{ and } g_1 = \begin{pmatrix} \alpha & \alpha + 1 \\ 1 & 1 \end{pmatrix}.$$

By construction, multiplication with one of these matrices can be implemented quite efficiently, making this hash function a rather interesting proposal. Unfortunately, more than a decade after the hash function was published, Grassl et al. demonstrated how a known result about a worst-case behavior of the Euclidean algorithm in the polynomial ring $\mathbb{F}_2[x]$ can be exploited to construct short collisions for this hash function [GIMS11]. Subsequent work by Petit and Quisquater provided a powerful technique to find preimages [PQ10], and this particular construction for a Cayley hash function is now considered insecure. However, this does not rule out the existence of better instances of the general idea, and in view of the elegance of the basic construction, the question for more secure parameters has some appeal to it. Petit and Quisquater [PQ13] suggest a challenge with parameters that are remarkably similar to the above-mentioned proposal which has been attacked successfully. Using again the group $\mathrm{SL}_2(\mathbb{F}_{2^m})$ and keeping the above notation, their challenge parameters can be expressed as

$$g_0 = \begin{pmatrix} \alpha^3 & 1 \\ 1 & 0 \end{pmatrix} \text{ and } g_1 = \begin{pmatrix} \alpha & \alpha + 1 \\ 1 & 1 \end{pmatrix}.$$

The subtle change in the generator g_0 appears to be a non-trivial obstacle for the earlier techniques that were used to construct collisions and preimages in the group $SL_2(\mathbb{F}_{2^m})$.

8.4   Message authentication codes

As already clear from its name, a message authentication code (MAC) is concerned with ensuring the authenticity of a message rather than ensuring confidentiality. It has some similarity with digital signatures as discussed in Chapter 10, but a major difference is the assumption that sender and receiver have a secret key in common. Unlike digital signatures, MACs are not intended to provide a *publicly* verifiable form of authentication. Both creating and verifying a tag involves a secret key.

DEFINITION 8.2 [Message Authentication Code]   *A message authentication code is given by a triple of polynomial time algorithms* $(\mathcal{K}, \mathcal{T}, \mathcal{V})$ *along with a family of message spaces* $\{M_\ell\}_{\ell \in \mathbb{N}}$ *as follows.*

$\mathcal{K}$: *a probabilistic key generation algorithm that on input a security parameter* 1^ℓ *outputs a secret key sk.*

$\mathcal{T}$: *a probabilistic tagging algorithm that on input a secret key sk and a message* $m \in M_\ell$ *outputs a message tag t or a dedicated error symbol* $\perp$.

$\mathcal{V}$: *a deterministic verification algorithm that on input a secret key sk and a message* $m \in M_\ell$ *outputs* VALID *or* INVALID, *indicating that a tag is considered as valid or invalid.*

To ensure correctness, we require that for a key sk generated by the key generation algorithm $\mathcal{K}$, *the event*

$$\exists m \in M_\ell : \mathcal{V}_{sk}(\mathcal{T}_{sk}(m)) \neq \text{VALID}$$

occurs with negligible probability only.

In a typical case, the message space consists of arbitrary bitstrings of bounded length, the key generation is a simple uniform random choice in $\mathbb{F}_2^\ell$, the tagging algorithm $\mathcal{T}$ is deterministic, and executing $\mathcal{V}$ amounts to recomputing the tag produced by $\mathcal{T}$. In fact, similarly as for block ciphers, for practical schemes, it is common to restrict to a small set of choices for the key length. The standard security notion for a message authentication scheme is existential unforgeability against adaptive chosen message attack, which is defined analogously in the case for signatures (see Definition 10.2). The idea is that even having message tags for chosen messages available, it remains infeasible to produce a valid tag for a new message. When formalizing this, one grants the (polynomial time) adversary access to a tagging oracle and a verification oracle, both of which have access to the secret key to compute and verify message tags, respectively. The adversary can query these oracles in an adaptive manner and to win the game has to create—with non-negligible success probability—an existential forgery, i. e., a message along with a valid tag such that this message has not been submitted to the tagging oracle yet.

If we are willing to model a cryptographic hash function as a random oracle $H : \{0,1\}^* \longrightarrow \{0,1\}^\ell$, then there is a natural candidate construction to derive a message authentication code from H. Namely, the key generation simply selects a secret bitstring $sk \in \{0,1\}^\ell$, and to compute the message tag for a message $m \in \{0,1\}^*$, we compute $\mathcal{T}_{sk}(m) = H(sk\|m)$. The corresponding verification algorithm recomputes the message tag and compares the result with the tag that needs to be verified. In the random oracle model, this construction is secure in the sense of existential unforgeability under adaptive chosen message attack. Nonetheless, this construction is avoided in practice, when a cryptographic hash function is used instead of H—Exercise 61 illustrates why this construction is problematic. In the next two sections we take a look at two "real-world" message authentication codes—HMAC and CMAC.

8.4.1   Keyed-Hash Message Authentication Code

The design of the *Keyed-Hash Message Authentication Code (HMAC)* as described by Krawczyk et al. [KBC97] and in [NIS08] is well suited for hash functions that follow the Merkle–Damgård design. In fact, the specification in [KBC97] notes that the underlying hash function is assumed *to be a cryptographic hash function where data is hashed by iterating a basic compression function on blocks of data*. We write $h : \{0,1\}^* \longrightarrow \{0,1\}^\ell$ for the cryptographic hash function that is used.[1] The length n of the input blocks accepted by the underlying compression function

$$f : \mathbb{F}_2^\ell \times \mathbb{F}_2^n \longrightarrow \mathbb{F}_2^\ell$$

is assumed to be a multiple of 8, and ℓ can be different from n. The secret key sk in HMAC is a random bitstring, e. g., chosen uniformly at random after fixing the desired key length. If the length of sk is greater than n, then sk is replaced with $h(sk)$.

In a first step of the tagging algorithm it is ensured that sk has length exactly n. If the length of sk is shorter, then we append 0s until we end up

---

[1]The domain of definition of h can also be a proper subset of $\{0,1\}^*$, e. g., because of applying Merkle–Damgård strengthening with a fixed-length encoding of the message length.

with an $sk \in \mathbb{F}_2^n$. By means of two key-independent constants we derive keys

$$\overbrace{}^{n/8 \text{ repetitions of } 0,0,1,1,0,1,1,0}$$

$$
\begin{aligned}
sk_{in} &= sk + (0,0,1,1,0,1,1,0,\ldots,0,0,1,1,0,1,1,0) \in \mathbb{F}_2^n \text{ and} \\
sk_{out} &= sk + (0,1,0,1,1,1,0,0,\ldots,0,1,0,1,1,1,0,0) \in \mathbb{F}_2^n
\end{aligned}
$$

$$\underbrace{}_{n/8 \text{ repetitions of } 0,1,0,1,1,1,0,0}$$

from sk. With these two keys, the message tag for $m \in \{0,1\}^*$ computes to

$$h(sk_{out} \| h(sk_{in} \| m)),$$

which is verified by recomputing the tag and checking for equality.

At first glance, tagging a message and verifying a tag with HMAC each require two applications of the underlying cryptographic hash function. It is important to note, however, that the outer call to the hash function involves only a short input.

8.4.2   Cipher-based Message Authentication Code

The *Cipher-based Message Authentication Code (CMAC)* is described in [Dwo05] and can be seen as mode of operation of a block cipher. We write n for the corresponding block length, so for AES we have $n = 128$. Analogously as in Section 7.4 the result of applying the block cipher to a single n-bit block $m \in \mathbb{F}_2^n$ using secret key K, is represented as $\text{Enc}_K(m)$. We will not need the decryption operation of the block cipher for CMAC.

Let us start with the key generation algorithm. Basically, only a key K for the block cipher needs to be generated (uniformly) at random. From this key, two subkeys $K^{(1)}, K^{(2)} \in \mathbb{F}_2^n$ are derived, using a uniquely determined irreducible polynomial $r_n(x) \in \mathbb{F}_2[x]$ of degree n. Specifically $r_n(x)$ is the lexicographically smallest element in the set of all irreducible polynomials of degree n with a minimal number of non-zero coefficients.

Example 8.2
For $n = 128$ we have $r_{128}(x) = x^{128} + x^7 + x^2 + x + 1$.   ⬜

To determine $K^{(1)}$ and $K^{(2)}$, we encrypt an all-zero block with K to obtain $(\gamma_{n-1}, \ldots, \gamma_0) = \text{Enc}_K(0^n) \in \mathbb{F}_2^n$. We interpret the resulting ciphertext as a coefficient vector of an element in the field $\mathbb{F}_2[x]/(r_n(x))$, compute

$$
\begin{aligned}
k_{n-1}^{(1)} x^{n-1} + \cdots + k_0^{(1)} &= x \cdot (\gamma_{n-1} x^{n-1} + \cdots + \gamma_0) \pmod{r_n(x)} \text{ and} \\
k_{n-1}^{(2)} x^{n-1} + \cdots + k_0^{(2)} &= x^2 \cdot (\gamma_{n-1} x^{n-1} + \cdots + \gamma_0) \pmod{r_n(x)},
\end{aligned}
$$

and finally set $K^{(i)} = (k_{n-1}^{(i)}, \ldots, k_0^{(i)})$ for $i = 1, 2$.

The core of the message tag computation in CMAC is the CBC mode as discussed in Section 7.4.2, but the subkeys $K^{(1)}$ and $K^{(2)}$ are used for a special handling of the last block of the message. More precisely, to compute a tag on a message $m \in \{0,1\}^*$, we first rewrite m as a sequence $m = m_1||\ldots||m_{r-1}||m_r$ with $m_1,\ldots,m_{r-1} \in \mathbb{F}_2^n$ and $m_r \in \mathbb{F}_2^{n-\delta}$ with $\delta \in \{0,\ldots,n-1\}$.[2] We define a modified last block m_r^* as

$$m_r^* = \begin{cases} K^{(1)} + m_r & \text{, if the length } |m_r| \text{ of the last block is } r \\ K^{(2)} + (m_r||10^{n-|m_r|-1}) & \text{, otherwise.} \end{cases}$$

Now the block cipher is applied for encryption to $m_1,\ldots,m_{r-1},m_r^*$ in CBC mode, using the secret key K. All ciphertext blocks with exception of the last block, let us call it t, are discarded. If the desired output length for the tag is n bits, then t serves as tag for m. If the desired length for the tag is $n' < n$, the tag consists of the n' left-most bits of t. In either case, to test the validity of a tag, the verifier recomputes the tag using the secret key K and checks for equality.

8.5 Summary and further reading

In this section we learned about the basic properties expected from a cryptographic hash function. Next to dedicated constructions, it is possible to derive a hash function from a block cipher, but it is important to keep in mind the restrictions on the output length imposed by the birthday phenomenon. Cayley hash functions are an elegant attempt to build cryptographic hash functions within a group-theoretic framework. To protect the authenticity of data, message authentication codes can be used, and with HMAC and CMAC we have seen two popular constructions.

From the perspective of group theory, Cayley hash functions are an interesting idea to derive a hash function, but there is so far no specific parameter choice that is widely accepted as being secure and usable for practical applications. Dedicated constructions, in particular the SHA family [NIS12, NIS14], are much more popular, and they are designed to offer high performance on typical computer platforms. Block-cipher based constructions are popular as well, but for a block cipher like AES, where the output length is only 128 bit, the construction in Section 8.2 is problematic. In view of the birthday phenomenon it is natural to ask how one can (efficiently) derive a hash function with output length $2n$ when starting out with a block cipher of length n. A paper by Lee and Stijm from CT-RSA 2011 [LS11] suggests such a construction and can serve as a starting point for learning more about this line of research. To ensure authenticity and confidentiality of a message simultaneously, one would naturally combine a MAC with a symmetric encryption

---

[2]If m is the empty string, we set $r = 1$ and $m_r = m$.

scheme, e. g., in an Encrypt-then-MAC or MAC-then-Encrypt style. Bellare and Namprempre [BN08] explore provable guarantees of such constructions.

8.6   Exercises

Exercise 55 *Show that the probability (8.3) for finding a preimage converges to 0 if $k = c \cdot n^\alpha$ for some $c > 0$ and $0 < \alpha < 1$.*

Exercise 56 *Let $h : \{0,1\}^* \longrightarrow \{0,1\}^{256}$ be a hash function. In an attempt to formalize collision-resistance of h, we require that for each probabilistic polynomial time algorithm the probability to output, given 1^ℓ, two bitstrings $m_0, m_1 \in \{0,1\}^*$ with $m_0 \neq m_1$ and $h(m_0) = h(m_1)$ is negligible. Show that no such function h exists.*

Exercise 57 *Fix an arbitrary value h in the Davies–Meyer scheme. Find a message m_h that results in the output h of the compression function (8.4).*

Exercise 58 *Consider the symmetric group S_n, and let $S \subseteq S_n$ be a generating set for S_n, with elements in S being represented in cycle notation. Is the resulting Cayley hash function collision-resistant?*

Exercise 59 *Suppose we use the Tillich–Zémor parameters with a binary field $\mathbb{F}_{2^m}$ where $m \in \{130, \ldots, 170\}$. Show that if the adversary is given a hash value $h(m)$ of a bitstring $m \in \mathbb{F}_2^{128}$, the bitstring m is determined uniquely and can be recovered efficiently.*

Exercise 60 *Suppose that you are allowed to control the irreducible polynomial $f(x) \in \mathbb{F}_2[x]$ used in the Tillich–Zémor hash function to represent the underlying finite field $\mathbb{F}_2[x]/(f(x))$. The only restrictions are that $f(x)$ is irreducible and its degree must be a prime number between 130 and 170. Can you fix $f(x)$ in such a way that you know a collision of the resulting hash function?*

Exercise 61 *Consider the message authentication code described at the beginning of Section 8.4.1 using the tagging algorithm $\mathcal{T}_{sk}(m) = H(sk||m)$ with a random oracle H.*

(a) Show that this message authentication code is existentially unforgeable under adaptive chosen message attacks.

(b) Suppose we replace H with a hash function as described in Algorithm 8.1, using the block length $n = 256$ for the compression function. To pad

inputs of the hash function in such a way that the message length is a multiple of 256, *we apply the padding in Example 8.1 with* $\gamma = 128$. *Suppose that with this parameter choice the adversary learns that for some secret key* $sk \in \mathbb{F}_2^{128}$ *the message tag* $\mathcal{T}_{sk}(001)$ *for* 001 *is the all-zero bitstring. Find a message* $m \neq 001$ *along with* $\mathcal{T}_{sk}(m)$.

Exercise 62 *Can you see a motivation why the keys* sk_{in} *and* sk_{out} *in HMAC are chosen to be of length* n, *even if the original secret key* sk *is shorter?*

Exercise 63 *Just like CMAC, the* Cipher Block Chaining MAC (CBC-MAC) *builds on a block cipher. The secret key of the MAC is a secret key* K *of the underlying block cipher. In the basic version[3] of CBC-MAC, to compute the tag for a message* m, *we proceed as follows, where* n *is the length of the underlying block cipher:*

1. *Apply the padding from Example 8.1 to ensure that the length of* m *is an integer multiple of* n.

2. *Using the all-zero vector as initialization vector, encrypt* m *with the block cipher in CBC mode. The last ciphertext block is the (n-bit) tag for* m — *all other ciphertext blocks are discarded.*

The verifier simply recomputes the tag with the secret key and checks for equality with the tag in question.

Suppose we use the above scheme with $n = 256$ *and parameter* $\gamma = 64$ *for the binary length encoding in the padding scheme in Example 8.1. Show that the resulting message authentication scheme is not secure in the sense of existential unforgeability under adaptive chosen message attack.*

---

[3] As explained in [MvOV96, Algorithm 9.58], one can add a post-processing step.

Part IV

Other Cryptographic Constructions

Chapter 9

Key establishment protocols

9.1 Setting the stage

Key establishment (also called key exchange) protocols are cryptographic schemes allowing a set of users connected through an insecure communication network to agree on a common secret value. This value is typically a secret key that can subsequently be used by the set of participants for a variety of cryptographic purposes. When attempting to design a key establishment protocol the first relevant issues to consider are:

- *the number of involved entities:* protocols involving only two users are simpler than those designed for larger groups. In particular, in larger groups a subset of legitimate participants may act cooperatively in a malicious way, or users may join/leave the protocol at different steps raising efficiency and security concerns;

- *the influence different participants have on the output key:* key transport schemes aim at the secure transmission of a key typically generated by a fully trusted entity involved in the protocol. On the other hand, key agreement protocols are designed in a *cooperative* way, in the sense that all participants have some influence in the final output value;

- *the way legitimate participants authenticate themselves:* it may be the case that the communication network is authentic or users may authenticate themselves as individuals—e.g., using digital signatures—or as group members—using passwords.

We have already seen a first example of a key exchange protocol, the Diffie–Hellman key agreement from Section 3.1.2. Note that this is a two-party protocol in which implicitly it is assumed that the communication channel is authenticated. Indeed, if this is not the case, a third (malicious) participant Eve may very easily mount a *middleperson attack*; impersonating Bob to Alice and Alice to Bob. As a result, Alice and Bob end up with different keying material, which is also owned by Eve who may interfere in their future communication at will. Denoting by G a public cyclic group of order q and g a public generator of G, this situation is described in Figure 9.1.

Alice	Eve	Bob
$a \leftarrow \mathbb{Z}_q$	$e \leftarrow \mathbb{Z}_q$	$b \leftarrow \mathbb{Z}_q$

$$h_1 = g^a \longrightarrow$$

$$\longleftarrow h_2 = g^b$$

$$\check{h}_1 = g^e$$

$$\check{h}_2 = g^e$$

$$K_{AB} = g^{ae} \qquad K_{EB} = g^{be} \qquad K_{AB} = g^{eb}$$

$$K_{EA} = g^{ae}$$

FIGURE 9.1: Middleperson attack on Diffie–Hellman key agreement.

Let us proceed with our informal analysis of the Diffie–Hellman key agreement. Once we assume the communication channel to be authenticated, we may take for granted that the key exchange is correct, namely, both Alice and Bob will end up sharing the same group element g^{ab}. Actually, checking a key exchange protocol for correctness is a bit more involved, for if the set of participants is larger than 2 we should guarantee that everyone knows with whom one is sharing a key, and if concurrent executions are allowed, then keys should also be "tagged" consistently. We will make all this more precise later. Further, it remains to argue that the exchanged information is sufficiently *unknown* to others to be used as a secret key. This is formalized via an *indistinguishability* property similar to the one we introduced when defining IND-CPA security (see Definition 4.4). The idea is that any external observer should not be able to tell the output value (g^{ab}) apart from a truly random value drawn from G. Thus, the requirement we need to prove security of the above scheme is the Decisional Diffie–Hellman assumption we discussed back in Section 4.2 (see Assumption 4.1).

One nice feature of the Diffie–Hellman scheme is that it can be adapted to be used in larger groups; the first to derive a group key establishment building on Diffie–Hellman key agreement were M. Burmester and Y. Desmedt [BD95]. Further refinements of this construction were later designed to attain very strong guarantees even in the presence of malicious insiders.

We summarize the fundamentals of the Burmester–Desmedt scheme for establishing a key within a set of users $\{U_1, \ldots, U_n\}$, where indices are to be taken in a cycle. For this protocol, we assume arbitrary point-to-point connections among participants to be available, and a *broadcast* is understood as simultaneous point-to-point delivery of messages to all intended recipients. The initial setting is that of standard two-party Diffie–Hellman; a cyclic group G of prime order q and a generator g of G are made public. In the first round the participants choose uniformly at random a private exponent $r_i \in \mathbb{Z}_q$, broadcast $z_i = g^{r_i}$ and compute a Diffie–Hellman key with their neighbors. Subsequently, each participant computes the quotient of the keys agreed with neighboring participants $X_i = (z_{i+1}/z_{i-1})^{r_i}$ and broadcasts this value. It is as

Round 1:

> **Computation:** Each U_i chooses $r_i \leftarrow \mathbb{Z}_q$, computes $z_i = g^{r_i}$.
>
> **Broadcast:** Each U_i broadcasts z_i.

Round 2:

> **Computation:** Each U_i computes $X_i = (z_{i+1}/z_{i-1})^{r_i}$.
>
> **Broadcast:** Each U_i broadcasts X_i.
>
> **Key computation:** Each U_i computes the session key
>
> $$\mathsf{sk}_i^{s_i} = (z_{i-1})^{nr_i} \cdot X_i^{n-1} \cdot X_{i+1}^{n-2} \cdots X_{i+n-2}.$$

FIGURE 9.2: Burmester–Desmedt group key establishment.

a result possible for each U_i to compute a session key $\mathsf{sk}_i = (z_{i-1})^{nr_i} \cdot X_i^{n-1} \cdot X_{i+1}^{n-2} \cdots X_{i+n-2}$, which is actually the product of all two-party keys that have been exchanged along the "cycle." A detailed overview of the exchanged messages is given in Figure 9.2.

The above scheme can be proved secure in an established model if only passive adversaries are taken into account. In terms of efficiency it can be improved in that there exist provable secure key exchange protocols which require less communication. Few subsequent constructions for key exchange make use of mathematical tools that in an essential manner deviate from Diffie–Hellman like assumptions in different cyclic groups, and we will see some examples. For now, let us see a construction (due to Pieprzyk and Li) of a group key agreement scheme based on *secret sharing* techniques.

Example 9.1 [Pieprzyk and Li's group key agreement]
Let G be a cyclic group of prime order q, g a generator of G, and denote by $\{U_1, \ldots, U_m\}$ the set of participants. Before any key is established, each participant U_i engages in an $(m + 1, 2m)$-secret sharing scheme in $\mathbb{Z}_q^*$ with every other participant. Namely, U_i chooses a polynomial

$$p_i(x) = a_{i,0} + a_{i,1}x + \cdots + a_{i,m}z^m$$

where each coefficient is selected uniformly at random from $\mathbb{Z}_q^*$. Further, U_i evaluates this polynomial at $2m$ points, yielding the sets:

$$\text{SecSet}_i = \{p_i(2j - 1) \mid j = 1, \ldots, m\} \text{ and}$$
$$\text{SharSet}_i = \{p_i(2j) \mid j = 1, \ldots, m\}.$$

While the first set is kept secret by U_i, each participant U_j is given the element $p_i(2j)$ from the second set. Finally, define $P(x) = p_1(x) + \cdots + p_m(x)$. Once

---

Round 0:

   Computation T selects $r \leftarrow \mathbb{Z}_q^*$

   Broadcast T broadcasts $\alpha = g^r$ to each U_i, where g is some generator of $\mathbb{Z}_q^*$

Round 1:

   Computation Each U_i computes $\beta_{ij} = \alpha^{p_i(2j-1)}$ for $j = 1, \ldots, m$.

   Broadcast Each U_i broadcasts β_{ij}, for $j = 1, \ldots, m$.

   Key computation: Each U_i computes

$$\alpha^{P(2j-1)} = \Pi_{i=1}^m \beta_{ij} \text{ for } j = 1, \ldots, m \text{ and}$$
$$\alpha^{P(2i)} = \alpha^{\sum_{j=1}^m p_j(2i)}.$$

   Now, using Lagrange interpolation in the exponents, U_i computes the shared key $K = \alpha^{P(0)}$.

FIGURE 9.3:   Pieprzyk and Li group key agreement.

this initialization is completed, different key agreement executions may proceed, with the help of a trusted party T (that may be one of the participants) as described in Figure 9.3.[1]

$\Box$

9.1.1   Provable security for key exchange protocols

Writing formal proofs of security for key exchange protocols is a rather involved task. As it was the case for public-key encryption schemes, the different formal models of security for key exchange have been motivated by existing attacks on published schemes, and many protocols have been proposed (and widely used) without a rigorous security analysis. Currently, the most deployed security models for key exchange protocols build on the model of Bellare and Rogaway [BR94a]. We here summarize the basic notions involved in this model, in an attempt to formalize the desirable properties of a key exchange protocol, in particular when it involves more than two participants.

Participants.   The (potential) protocol participants are modeled as a finite set $\mathcal{U}$ of fixed size with each U_i being a probabilistic polynomial time Turing machine. Each protocol participant $U_i \in \mathcal{U}$ may execute a polynomial number

---

[1]In the protocol description from the authors [PL02], Round 0 was actually omitted and formulated by means of a public register (R).

of protocol instances in parallel. We will refer to instance s_i of participant U_i as *oracle* $\Pi_i^{s_i}$ ($i \in \mathbb{N}$). Each such oracle may be taken for a process executed by U_i and has assigned seven variables $\mathsf{state}_i^{s_i}$, $\mathsf{sid}_i^{s_i}$, $\mathsf{pid}_i^{s_i}$, $\mathsf{sk}_i^{s_i}$, $\mathsf{term}_i^{s_i}$, $\mathsf{used}_i^{s_i}$ and $\mathsf{acc}_i^{s_i}$:

$\mathsf{used}_i^{s_i}$ indicates whether this oracle is or has been used for a protocol run. The $\mathsf{used}_i^{s_i}$ flag can only be set through a protocol message received by the oracle due to a call to the Execute-oracle or a call to the Send-oracle (see below);

$\mathsf{state}_i^{s_i}$ keeps the state information during the protocol execution;

$\mathsf{term}_i^{s_i}$ shows if the execution has terminated;

$\mathsf{sid}_i^{s_i}$ denotes a (non-secret) session identifier that can serve as identifier for the session key $\mathsf{sk}_i^{s_i}$;

$\mathsf{pid}_i^{s_i}$ stores the set of identities of those participants that $\Pi_i^{s_i}$ aims at establishing a key with—including U_i;

$\mathsf{acc}_i^{s_i}$ indicates if the protocol instance was successful, i.e., the participant accepted the session key;

$\mathsf{sk}_i^{s_i}$ stores the session key once it is accepted by the oracle $\Pi_i^{s_i}$. Before acceptance, it stores a distinguished NULL value.

We suppose that an oracle $\Pi_i^{s_i}$ must accept the session key constructed at the end of the corresponding protocol instance if no deviation from the protocol specification occurs.

Initialization. Before the actual key establishment protocol is executed for the first time, we allow an initialization phase to take place. Typically, at this a public key/secret key pair (SK_i, PK_i) is generated for each participant $U_i \in \mathcal{U}$—this may, for instance, be used for authentication. At this point, SK_i is revealed to U_i only, and PK_i is given to all participants. Even an active adversary (see adversarial model below) is not given the ability to influence the initialization phase.

Communication network. We assume arbitrary point-to-point connections among the participants to be available. As connections are potentially under adversarial control, the network is non-private and fully asynchronous.

Adversarial model. For a passive adversary $\mathcal{A}$ all messages sent by protocol participants are sent as specified in the protocol description, but may be eavesdropped by $\mathcal{A}$. An active adversary $\mathcal{A}$ has full control of the communication network and may delay, suppress, and insert messages at will. To make the adversary's capabilities explicit, the subsequently listed oracles are used. An active adversary is a probabilistic polynomial time Turing machine that may execute any of these, whereas a passive adversary is a probabilistic polynomial time Turing machine that is only given access to the Execute, Reveal, and Test oracles.

Execute($\{U_1, U_2, \ldots, U_r\}$) This executes the protocol among unused instances $\Pi_i^{s_i}$ of the specified parties and returns a transcript of the protocol run (listing all messages sent during the protocol execution among the oracles $\Pi_i^{s_i}$).

Send(U_i, s_i, M) This sends the message M to the instance $\Pi_i^{s_i}$ and outputs the reply generated by this instance. If the adversary calls this oracle with an unused instance $\Pi_i^{s_i}$ and $M = \{U_1, \ldots, U_r\}$, then $\Pi_i^{s_i}$'s $\mathsf{pid}_i^{s_i}$-value is initialized to the value $\mathsf{pid}_i^{s_i} = M$ and the $\mathsf{used}_i^{s_i}$-flag is set. If the oracle $\Pi_i^{s_i}$ sends a message in the protocol right after receiving M, then Send returns this message to the adversary.

Reveal(U_i, s_i) yields the session key $\mathsf{sk}_i^{s_i}$ and the session identifier $\mathsf{sid}_i^{s_i}$.

Corrupt(U_i) reveals the long-term secret key SK_i (if any) of U_i to the adversary. Given a concrete protocol run, involving oracles $\Pi_i^{s_i}$ of participants $U_1, \ldots, U_k$ we say that participant $U_{i_0} \in \{U_1, \ldots, U_k\}$ is *honest* if and only if no query of the form Corrupt(U_{i_0}) has been made by the adversary.

Test(U_i, s_i) Only one query of this form is allowed for an active adversary $\mathcal{A}$. Provided that $\mathsf{sk}_i^{s_i}$ is defined, (i.e., $\mathsf{acc}_i^{s_i} = \mathsf{true}$ and $\mathsf{sk}_i^{s_i} \neq \mathrm{NULL}$), $\mathcal{A}$ can execute this oracle query at any time when being activated. Then with probability $1/2$ the session key $\mathsf{sk}_i^{s_i}$ and with probability $1/2$ a uniformly at random chosen element from the session key space is returned.

REMARK 9.1 In the definition above, the Corrupt oracle allows to expose long-term secrets and is used for addressing so-called *forward secrecy*—see Exercise 65—as well as for modeling malicious protocol participants. ∐

Correctness. This property basically expresses that the protocol will establish a good key without adversarial interference and allows us to exclude "useless" protocols. We take a group key establishment protocol for *correct* if in the presence of a passive adversary indeed a common key along with a common identifier is established:

DEFINITION 9.1 [Correct group key establishment] *A group key establishment protocol* P *is called* correct *if in the presence of a passive adversary a single execution of the protocol for establishing a key among* $U_1, \ldots, U_r$ *involves* r *oracles* $\Pi_1^{s_1}, \ldots, \Pi_r^{s_r}$ *and ensures that with overwhelming probability all oracles:*

- *accept, i. e.,* $\mathsf{acc}_1^{s_1} = \cdots = \mathsf{acc}_r^{s_r} = \mathsf{true}.$

- *obtain a common session identifier* $\mathsf{sid}_1^{s_1} = \cdots = \mathsf{sid}_r^{s_r}$ *which is globally unique.*

- *have accepted the same session key* $\mathsf{sk}_1^{s_1} = \cdots = \mathsf{sk}_r^{s_r} \neq \mathrm{NULL}$ *associated with the common session identifier* $\mathsf{sid}_1^{s_1}.$

- *know their partners* $\mathsf{pid}_1^{s_1} = \mathsf{pid}_2^{s_2} = \cdots = \mathsf{pid}_r^{s_r}$ *and it is* $\mathsf{pid}_1^{s_1} = \{U_1, \ldots U_r\}.$

Partnering. For detailing the security definition, we will have to specify under which conditions a Test-query may be executed. To do so we fix the following notion of partnering.

DEFINITION 9.2 [Partnering] *Two oracles* $\Pi_i^{s_i}, \Pi_j^{s_j}$ *are partnered if* $\mathsf{sid}_i^{s_i} = \mathsf{sid}_j^{s_j}$, $\mathsf{acc}_i^{s_i} = \mathsf{acc}_j^{s_j} = \mathsf{true}$, *and both* $U_j \in \mathsf{pid}_i^{s_i}$ *and* $U_i \in \mathsf{pid}_j^{s_j}$.

Freshness. A Test-query should only be allowed to those oracles holding a key that is not for trivial reasons known to the adversary. An instance $\Pi_i^{s_i}$ is called *fresh* if none of the following two conditions hold:

- For some $U_j \in \mathsf{pid}_i^{s_i}$ a Corrupt(U_j) query was executed before a query of the form Send$(U_k, s_k, *)$ has taken place where $U_k \in \mathsf{pid}_i^{s_i}$.

- $\mathcal{A}$ queried Reveal(U_j, s_j) with $\Pi_i^{s_i}$ and $\Pi_j^{s_j}$ being partnered.

The idea here is that revealing a session key from an oracle $\Pi_i^{s_i}$ trivially yields the session key of all oracles partnered with $\Pi_i^{s_i}$, and hence this kind of "attack" will be excluded in the security definition.

Security. As a function of the security parameter ℓ we define the advantage $\mathbf{Adv}_{\mathcal{A}}(\ell)$ of an adversary $\mathcal{A}$ in attacking protocol P as

$$\mathbf{Adv}_{\mathcal{A}} = |2 \cdot \mathsf{Succ} - 1|$$

where Succ is the probability that the adversary queries Test on a fresh instance $\Pi_i^{s_i}$ and guesses correctly the bit b used by the Test oracle in a moment when $\Pi_i^{s_i}$ is still fresh.

DEFINITION 9.3 [Secure group key establishment] *We call the group key establishment protocol* P *secure if for any adversary* $\mathcal{A}$ *the function* $\mathbf{Adv}_{\mathcal{A}} = \mathbf{Adv}_{\mathcal{A}}(\ell)$ *is negligible.*

9.1.2 A secure construction

Let us now depict what a secure construction according to the model described in the previous section looks like. For this, we need to introduce a *computational* version of the Decisional Diffie–Hellman assumption we saw in 4.1

Assumption 9.1 (Computational Diffie–Hellman assumption)
Let $\{G_\ell\}_{\ell \in \mathbb{N}}$ be a family of, multiplicatively written, finite cyclic groups along with a polynomial time algorithm $\mathcal{G}$ that on input 1^ℓ outputs a generator g_ℓ for G_ℓ (along with polynomial time algorithms for computing the product and for deciding equality of elements in G_ℓ).

The Computational Diffie–Hellman (CDH) assumption states that every probabilistic polynomial time algorithm $\mathcal{A}$ given as input $g \leftarrow \mathcal{G}(1^\ell), g^\alpha$ and g^β —with $\alpha \leftarrow \{1, \dots, \mathrm{ord}(g)\}$ and $\beta \leftarrow \{1, \dots, \mathrm{ord}(g)\}$—has negligible probability in ℓ of computing $g^{\alpha\beta}$.

The protocol is depicted in Figure 9.4. It is a refinement of the Burmester–Desmedt group key establishment protocol we discussed in the previous section. Two-party keys are agreed on in a first communication round among neighboring participants; however, these are used for securing the communication and detecting messages that have been tampered with (replayed), rather than for constructing the session key. Messages are authenticated by means of a signature scheme, and the random oracle is used both to guarantee the session key indistinguishability and to bind the session identifier and the session key together. Assuming the CDH assumption holds for (G, g) (more precisely, the corresponding group family), the function $H(\cdot)$ is modeled as a random oracle with codomain $\{0, 1\}^\ell$ and the underlying signature scheme is secure in a strong sense,[2] then the protocol in Figure 9.4 is a secure group key agreement in the sense of Definition 9.3.

Proof sketch. Let ℓ be the security parameter. Let q_s and q_{ro} be polynomial bounds for the number of the adversary's queries to the Send respectively the random oracle. We begin by defining three events and give bounds for their probability that are negligible in ℓ.

Forge is the event that the adversary succeeds in forging an authenticated message $M_{U_i} \| \sigma_{U_i}$ for participant U_i without having queried Corrupt(U_i) and where M_{U_i} was not output by any of U_i's instances.

An adversary $\mathcal{A}$ that can reach Forge can be used for forging a signature for a given public key: This key is assigned to one of the n participants and $\mathcal{A}$ succeeds in the intended forgery with probability $\geq \frac{1}{n} \cdot \Pr[\mathsf{Forge}]$.

---

[2]the right notion is *existentially unforgeable under adaptive chosen message attacks* (see Definition 10.2)

Round 1:

Computation: Each U_i chooses $k_i \leftarrow \{0,1\}^\ell$, $x_i \leftarrow \mathbb{Z}_q^*$ and computes $y_i = g^{x_i}$.

U_n computes additionally $H(k_n)$.

Each U_i except U_n sets $M_i^{\mathrm{I}} = k_i \| y_i$.

U_n sets $M_n^{\mathrm{I}} = H(k_n) \| y_n$.

Further, each U_i computes a signature σ_i^{I} on $M_i^{\mathrm{I}} \| \mathsf{pid}_i^{s_i}$.

Broadcast: Each U_i broadcasts $(M_i^{\mathrm{I}} \| \sigma_i^{\mathrm{I}})$.

Check: Each U_i checks all signatures σ_j^{I} of incoming messages $(M_j^{\mathrm{I}} \| \sigma_j^{\mathrm{I}})$.

Round 2:

Computation: Each U_i computes $t_i^L = H(y_{i-1}^{x_i})$, $t_i^R = H(y_{i+1}^{x_i})$, $T_i = t_i^L \oplus t_i^R$ and $\mathsf{sid}_i^{s_i} = H(\mathsf{pid}_i^{s_i} \| k_1 \| \ldots \| k_{n-1} \| H(k_n))$.

Additionally, only U_n computes $k_n \oplus t_n^R$.

The participants $U_1, \ldots, U_{n-1}$ set $M_i^{\mathrm{II}} = \mathsf{sid}_i^{s_i} \| T_i$

U_n sets $M_n^{\mathrm{II}} = k_n \oplus t_n^R \| \mathsf{sid}_n^{s_n} \| T_n$.

Further, each U_i computes a signature σ_i^{II} of M_i^{II}.

Broadcast: Each U_i broadcasts $(M_i^{\mathrm{II}} \| \sigma_i^{\mathrm{II}})$.

Check: Each U_i checks all signatures σ_j^{II} of incoming messages. Then each U_i checks if

$$T_1 \oplus \cdots \oplus T_n = 0 \text{ and}$$
$$\mathsf{sid}_i^{s_i} = \mathsf{sid}_j^{s_j} \quad (j = 1, \ldots, n).$$

Moreover, each U_i $(i < n)$ checks the commitment $H(k_n)$ for k_n.

Key computation: Each participant U_i computes the session key $\mathsf{sk}_i^{s_i} = H(\mathsf{pid}_i^{s_i} \| k_1 \| \ldots \| k_n)$.

FIGURE 9.4:   A secure group key agreement protocol.

Thus, using $\mathcal{A}$ as black box we can derive an attacker defeating the existential unforgeability of the underlying signature scheme S with probability

$$\mathbf{Adv}_S^{\mathrm{cma}} \geq \tfrac{1}{n} \cdot \Pr[\mathsf{Forge}]$$
$$\Longleftrightarrow \Pr[\mathsf{Forge}] \leq n \cdot \mathbf{Adv}_S^{\mathrm{cma}}.$$

Here $\mathbf{Adv}_S^{\mathrm{cma}}$ denotes the advantage of the adversary in violating the existential unforgeability under an adaptive chosen message attack of the signature scheme, which is negligible by assumption. Thus, the event Forge occurs with negligible probability only.

$\mathsf{Collision}$ is the event that the random oracle produces a collision. A Send query causes at most 3 random oracle calls. Thus, the total number of random oracle queries is bounded by $3q_s + q_{ro}$ and the probability that a collision of the random oracle occurs is

$$\Pr[\mathsf{Collision}] \leq \frac{(3q_s + q_{ro})^2}{2^\ell},$$

which is negligible in ℓ.

Repeat is the event that an uncorrupted participant chooses a nonce k_i that was previously used by an oracle of some participant. There are at most q_s used instances that may have chosen a nonce k_i and thus Repeat happens with a probability

$$\Pr[\mathsf{Repeat}] \leq \frac{q_s^2}{2^\ell},$$

again negligible in ℓ.

To prove the security according to Definition 9.3, we consider a sequence of games. In these games we let the adversary $\mathcal{A}$ interact with a simulator, that in Game 0 offers the original protocol environment to $\mathcal{A}$, and subsequently we change the simulator's behavior in several small steps without affecting $\mathcal{A}$'s success probability significantly. Keeping track of the changes between subsequent games, in the last game we will be able to derive the desired negligible upper bound on $\mathbf{Adv}_{\mathcal{A}}$.

Game 0: In this game the protocol participants' instances are faithfully simulated for the adversary, i.e., the adversary's situation is the same as in the real model.

$$\mathbf{Adv}_{\mathcal{A}}^{\mathrm{Game}\ 0} = \mathbf{Adv}_{\mathcal{A}}.$$

Game 1: This game is aborted if one of the events Forge, $\mathsf{Collision}$, or Repeat occurs. Otherwise the game is identical with Game 0 and the adversary cannot detect the difference. Thus, for adversary $\mathcal{A}$'s advantage we have

$$|\mathbf{Adv}_{\mathcal{A}}^{\mathrm{Game}\ 1} - \mathbf{Adv}_{\mathcal{A}}^{\mathrm{Game}\ 0}| \leq \Pr[\mathsf{Forge}] + \Pr[\mathsf{Collision}] + \Pr[\mathsf{Repeat}].$$

Game 2: This game differs from Game 1 in the simulator's response in Round 2. If the simulator has to output the message of an instance $\Pi_i^{s_i}$ and none of the participants $U_\ell \in \mathsf{pid}_i^{s_i}$ is corrupted, then the simulator chooses random values from $\{0,1\}^\ell$ for $t_i^L = t_{i-1}^R$ and $t_i^R = t_{i+1}^L$ instead of querying the random oracle. To keep consistency, the same values have to be used in the neighbored instances subsequently.[3] By the random oracle assumption, the adversary can only detect the difference by querying the random oracle for $y_{i-1}^{x_i} = y_i^{x_{i-1}}$.

An adversary $\mathcal{A}$ that distinguishes Game 1 and Game 2 can be used as black box to solve a CDH instance. Two instances $\Pi_i^{s_i}$ and $\Pi_j^{s_j}$ are selected by randomly choosing two different users $U_i, U_j \in \mathcal{U}$ plus two numbers $s_i, s_j \in \{1, \ldots, q_s\}$. Game 2 only differs from Game 1, if at least one session is set up of uncorrupted users. To distinguish the games, the adversary has to query the random oracle with at least one Diffie–Hellman key, established between neighbors in a session with uncorrupted participants. These randomly chosen instances will be those neighbored participants with probability no less than $1/(n \cdot q_s)^2$.

A given CDH instance (g^a, g^b) is then assigned to $\Pi_i^{s_i}$ and $\Pi_j^{s_j}$ such that these instances will use g^a respectively g^b as their message for the first round.

If at some point now $\Pi_i^{s_i}$ and $\Pi_j^{s_j}$ do not qualify any longer to be neighbored participants in a session with only uncorrupted users, the simulation is aborted (as noted, this happens with probability bounded by $1/(n \cdot q_s)^2$).

Then a random index $z \in \{1, \ldots, q_{ro}\}$ is chosen and the adversary's z^{th} query to the random oracle is taken for the answer to the CDH challenge. The answer to the CDH challenge is correct if $\mathcal{A}$ distinguished the games with the chosen instances and also z was determined correctly. So we have

$$|\mathbf{Adv}_{\mathcal{A}}^{\text{Game 2}} - \mathbf{Adv}_{\mathcal{A}}^{\text{Game 1}}| \le \mathsf{Succ}_{(G,g)}^{\text{CDH}} \cdot q_{ro} \cdot n^2 \cdot q_s^2,$$

where $\mathsf{Succ}_{(G,g)}$ is a—under the CDH assumption—negligible upper bound for the success probability of the above algorithm to solve CDH.

Game 3: In this game the simulator changes the computation of the session key. Having received all messages of Round 2 for an instance $\Pi_i^{s_i}$, the simulator checks if all $U_j \in \mathsf{pid}_i^{s_i}$ are uncorrupted. If so, then the simulator chooses a session key $\mathsf{sk}_i^{s_i} \in \{0,1\}^\ell$ at random instead of querying the random oracle. For consistency the simulator will assign the same key to all partnered instances.

The only way for the adversary to detect the difference is by querying the random oracle for $H(\mathsf{pid}_i^{s_i}\|k_1\|\ldots\|k_n)$. However, k_n is information-theoretically unknown to the adversary. Thus, the adversary can only guess a random

---

[3]In the first round, the participants include the pid in the signature and thus know afterward that they obtained the same value pid. Hence, being neighbored is well defined.

value for k_n and query the random oracle at most q_{ro} times. This results in

$$|\mathbf{Adv}_{\mathcal{A}}^{\text{Game 3}} - \mathbf{Adv}_{\mathcal{A}}^{\text{Game 2}}| \leq \frac{q_{ro}}{2^\ell}.$$

None of the partners of the adversary's Test-instance are allowed to be corrupted or to be revealed (see Definition 9.3). Thereby, those instances were affected in Game 3 and use a random value as session key. Therefore, the adversary has only a probability of $1/2$ for guessing the bit of Test, yielding

$$\mathbf{Adv}_{\mathcal{A}}^{\text{Game 3}} = 0.$$

Putting the probabilities together we recognize the adversary's advantage in the real model as negligible:

$$\mathbf{Adv}_{\mathcal{A}} \leq \Pr[\mathsf{Forge}] + \Pr[\mathsf{Collision}] + \Pr[\mathsf{Repeat}] +$$
$$\mathsf{Succ}_{(\mathbb{G},g)}^{\text{CDH}} \cdot q_{ro} \cdot n^2 \cdot q_s^2 + \frac{q_{ro}}{2^\ell}.$$

The above proof sketch indicates that proving the security of a group key exchange protocol can be a rather involved task. Even though things are somewhat simpler in a two-party scenario, there are still many issues to address in order to achieve a secure and efficient design. In the next sections we focus our attention on several proposals for deriving two-party key exchange protocols from group-theoretic problems. These constructions were made at a rather conceptual level, considering essentially a stand-alone setting where no active adversaries are present. Still, they help us realize that even in this simplified scenario very strong computational guarantees are vital if the resulting keying material is to be trusted.

9.2   Anshel–Anshel–Goldfeld key exchange

Anshel et al. presented at the end of the 20th century a general method for deriving two-party key exchange protocols from non-Abelian algebraic structures [AAG99]. Their proposal was rather conceptual, yet it motivated a large number or researchers to look into the subject both for finding possible instantiations as well as for identifying vulnerabilities.

Here is an outline of the key exchange protocol—we do not give it in full generality but rather focus on the group-theoretic version. Assume a non-Abelian group G is publicly known, as well as two sets of elements $\{a_1, \ldots, a_k\}$, $\{b_1, \ldots, b_t\}$ from G. Alice and Bob choose reduced words (see Definition 6.1) in $\{a_1, \ldots, a_k\}$ and $\{b_1, \ldots, b_t\}$, respectively, yielding elements a and b. Further, Alice sends Bob the conjugated elements $b_1^a, \ldots, b_t^a$, while Bob sends Alice the corresponding $a_1^b, \ldots, a_k^b$. Now both Alice and Bob may compute the group element

$$a^{-1}b^{-1}ab = (b^a)^{-1}b = a^{-1}a^b.$$

Note that Alice computes a^b just by repeating the steps she followed for building the reduced word a from the "alphabet" $a_1^b, \ldots a_k^b$. This description is represented graphically in Figure 9.5, where the rewriting process for deriving a reduced word from a set/alphabet $A \subseteq G$ is denoted $w \sim_{rw} A$.

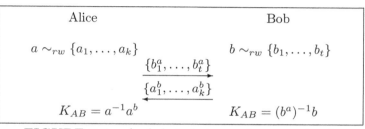

FIGURE 9.5: Anshel et al. key exchange protocol.

The authors suggested using braid groups as a suitable platform for realizing this scheme. This, and the proposal we will explore in Section 9.3 encouraged the work of many cryptanalysts which relatively soon came up with heuristic (yet damaging enough) attacks in the braid scenario—cf. our discussion in Section 6.3.2. However, only recently a polynomial time algorithm has been proposed by B. Tsaban for the underlying computational problem of retrieving the shared element K_{AB} from the public information [Tsa13], which is known as the *commutator KEP problem*.

Let us take a look at Tsaban's strategy. First of all, we may represent each element $b \in B_n$, through its Garside normal form $\Delta_n^i p_1 \cdots p_r$ (see Theorem 6.1) respectively by the tuple

$$(i, \mathbf{p})$$

where $\mathbf{p} = (p_1, \ldots, p_r)$. We can take $\mathbf{p}$ for a sequence of elements in S_n. The integer i is typically referred to as the *infimum* of b and denoted by $\inf(b)$. Similarly, r, the number of permutations involved in the normal form of b, is denoted by $\ell(b)$.

First, it is shown that without loss of generality, we may assume that the public braids $a_1, \ldots, a_k, b_1, \ldots, b_t$ all have infimum in $\{0, 1\}$. This simple observation yields that

$$-4m(l+1) \leq \inf(K_{AB}) \leq \inf(K_{AB}) + \ell(K_{AB}) \leq 4m(l+1),$$

where both a and b have been constructed multiplying at most m public generators (or inverses of them), and l is an upper bound on

$$\{\ell(a_1), \ldots \ell(a_k), \ell(b_1), \ldots, \ell(b_t)\}.$$

Define $M = 4m(l+1)$. Now the idea is to transfer the problem to a linear group, namely, to $\mathrm{GL}_n(p^d)$ where p is a prime larger than $2^{n^2 M}$ and d is larger than $2M$ (but both as small as possible). To move back and forth from the linear group, the main tool is the so called Lawrence–Krammer representation [Kra02]

$$LK : B_n \longrightarrow \mathrm{GL}_{\binom{n}{2}}\left(\mathbb{Z}\left[t^{\pm 1}, \frac{1}{2}\right]\right)$$

and the corresponding polynomial time algorithms for computing and inverting it ([CJ03]). The bound M, however, allows to work in a more convenient group than $\mathrm{GL}_{\binom{n}{2}}(\mathbb{Z}[t^{\pm 1}, \frac{1}{2}])$, namely, $\mathrm{GL}_n(\mathbb{F})$, where $\mathbb{F}$ is a finite field. The trick behind this attack is to exploit both the group-theoretic and the vector space structures related to elements in play, that are in the end $n \times n$ matrices over $\mathbb{F}$.

In the sequel, we assume without loss of generality that the public sets of generators are equal, namely $t = k$, and that all elements are from the linear group $G \subseteq \mathrm{GL}_n(\mathbb{F})$. The attack is structured in two phases, which we present in Figures 9.6 and 9.7.

Input: $b_1, \ldots, b_k \in G$.

Computation:

    1. Compute a basis $S = \{s_1, \ldots, s_u\}$ for $C(b_1, \ldots, b_k)$ by solving the system of linear equations

$$b_i X = X b_i \text{ for } i = 1, \ldots k.$$

    2. Compute a basis for $C(C(b_1, \ldots, b_k))$ by solving the system of linear equations

$$s_i X = X s_i \text{ for } i = 1, \ldots u.$$

Output: A basis for the vector subspace $C(C(b_1, \ldots, b_k))$

FIGURE 9.6: Tsaban's algorithm: Phase 1.

Here, given a set $S \subseteq G$, $C(S)$ denotes the *centralizer* of S in G, that is, the set of elements from G that commute with every element in S. It is easy to see that for every $S \subseteq G$, $C(S)$ is a vector subspace of the space of $n \times n$ matrices over $\mathbb{F}$. Phase 1 is concerned with computing a basis for $C(C(b_1, \ldots, b_k))$, which is later used in Phase 2 to derive a Las Vegas algorithm for retrieving the exchanged secret key (namely, it gives the right answer whenever it terminates, which can be proved to happen with overwhelming probability).

Input: $a_1, \ldots, a_k, b_1, \ldots, b_k, b_1^a, \ldots, b_k^a, a_1^b, \ldots, a_k^b \in G$

Computation:

1. Solve the system of linear equations

$$b_i X = X b_i^a, i = 1, \ldots, k$$

   for X and pick random solutions until an invertible solution A is found.

2. Solve the system of linear equations

$$a_i Y = Y a_i^b, i = 1, \ldots, k$$

   subject to the linear constraint that

$$Y \in C(C(b_1, \ldots, b_k))$$

   for Y and pick random solutions until an invertible solution B is found.

Output: $A^{-1} B^{-1} AB$

FIGURE 9.7:   Tsaban's algorithm: Phase 2.

There are a number of steps we have not really justified in the above attack; (like, for instance, why should invertible elements be found in the Phase 2 computation). We refer the interested reader to the original paper for a more detailed discussion. Last, to prove that the procedure actually works it remains to see that the output element $A^{-1} B^{-1} AB$ is indeed K_{AB}. To do so, it suffices to note that Aa^{-1} commutes with B and further $a^B = a^b$. As a result, we obtain

$$A^{-1} B^{-1} AB = A^{-1} B^{-1} (Aa^{-1}) aB = a^{-1} B^{-1} aB = a^{-1} a^B = a^{-1} a^b = K_{AB}.$$

9.3   Braid-based key exchange

Almost simultaneous to Anshel et al.'s proposal we described in the previous
section, a Diffie–Hellman-like key establishment protocol using braid groups
as a base was proposed in [KLC$^+$00]. The encryption scheme we depicted in
Figure 6.8 is actually derived from this key exchange protocol, in the same
way ElGamal encryption (see Figure 4.4) was derived from the Diffie–Hellman
key establishment.

Let A, B be two commuting subgroups of the braid group B_n, for some suitable n (in the original paper, proposed as $\simeq 80$). The authors of this scheme
proposed to take $A = LB_n$ and $B = RB_n$, but any commuting subgroups
would actually do. The key exchange protocol is described in Figure 9.8,
where $g \in B_n$ is assumed to be public.

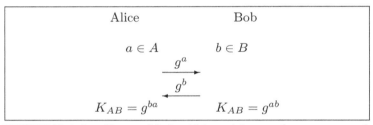

FIGURE 9.8:   Ko et al.'s braid-based key exchange protocol.

As it was originally presented, the scheme was soon identified to produce
weak keys, namely, the group element g^{ab} could be efficiently distinguished
from a "random" conjugate of g, and the output key should rather be defined by means of an ideal hash function (as it was actually specified in the
corresponding encryption scheme—see Figure 6.8). At CRYPTO 2001, Lee
et al. presented a paper with some first proposals for using braid groups for
pseudo-random number generation [LLH01]. We do not detail their proposal
in this book, but give a brief summary of the successful attack on it presented
at Eurocrypt 2002.[4]

The assumption on which the security of Lee et al.'s construction is based
is a variant of the braid Diffie–Hellman decision problem, which we can state
as follows—ignoring for the moment the question of what exactly we mean by
a random braid:

---

[4]The reader following the dates of the pertinent publications will probably notice that
the end of the 20$^\text{th}$ and beginning of this century was indeed a very active time for the
exploration of the cryptographic potential of braid groups.

Assumption 9.2 (Ko–Lee decisional assumption)

Let $a, b, c \in B_n$ such that for some (randomly chosen) $x \in LB_n$ and $y \in RB_n$ we have

$$b = x^{-1}ax \text{ and}$$

$$c = y^{-1}ay.$$

Then it is hard to distinguish the group element $d = a^{xy}$ from a random conjugate of a.

R. Gennaro and D. Micciancio proved this assumption to be false, by providing in [GM02] a polynomial time algorithm solving the corresponding decisional problem. Moreover, they also provided an explicit attack against the pseudo-random number generator proposed using that assumption. The goal of Gennaro and Micciancio's algorithm is to disprove the Ko–Lee decisional assumption by showing that for any choice of probability distribution on a, x, and y, it is always possible to derive some information about d from a, b, and c. Let us assume that n is even. Then Gennaro and Micciancio's strategy can be split into four main parts:

i. First we note that the permutation induced by x acts only on the leftmost letters of $\{1, \ldots, n\}$, and the permutation induced by y only on the corresponding right-most letters. Thus, the permutation induced by d must satisfy certain constraints. This allows the authors of the attack to argue that whenever a is not a pure braid, the Ko–Lee decisional assumption does not hold.

ii. Suppose a is a pure braid, then one can distinguish d from a random conjugate of a by comparing the two projections of d on $B_{\frac{n}{2}}$ (i.e., considering only the first or last $\frac{n}{2}$ strands of d) with the corresponding projections of b (first $\frac{n}{2}$ strands) and c (last $\frac{n}{2}$ strands). This will coincide significantly more often if d is a "Ko–Lee key" than if d is a random conjugate. This strategy is however not valid if both (left and right) projections of a yield the identity.

iii. Now assume a is pure, and the attack from ii. does not apply. They perform a similar comparison projecting now the public braids on different sets of $\frac{n}{2} + 1$ strands, which allows to distinguish d from random unless all such projections of a coincide.

iv. For the remaining choices of a, the authors argue that all conjugates involving braids in either LB_n or RB_n are trivial (i.e., coincide with a). As a consequence, the Ko–Lee decisional problem always has a "yes" answer.

Thus, to explore the security of the key exchange protocol of Ko et al. we go forward assuming a random oracle is introduced in the last step to circumvent the above analysis strategy for distinguishing a key from a random group element. Now, the underlying problem we are concerned with is a variant of the *braid Diffie–Hellman search problem*. Thus, the techniques we already mentioned when speaking of the *conjugacy problem* in the braid groups (see Section 6.3.2) offer a (quite successful) attack vector against this scheme. Moreover, the same linearization technique presented against the commutator KEP in the previous section refines the Cheon–Jun attack [CJ03] and yields a provable polynomial time algorithm to defeat the underlying computational problem. Again the problem can be reduced to the case when we consider two commuting subgroups A, B of the general linear group $GL_n(\mathbb{F})$, for $\mathbb{F}$ a suitable finite field. Now if B is made explicit by a generating set $\{b_1, \ldots, b_k\}$ then the key is retrieved by finding an invertible solution of the system

$$xg^a = gx$$
$$xb_1 = b_1 x$$
$$\vdots$$
$$xb_k = b_k x$$

of $(k+1)n^2$ linear equations in n^2 unknowns (see Exercise 66).

9.4 Constructions over matrix groups

In a sense, linear groups are "computationally friendly" which is a double-edged sword when it comes down to cryptographic applications. The fact that they can be handled both group-theoretically and from a linear algebra perspective appears (as we have already seen) more often a drawback than an advantage. In this section we review two key exchange protocols that have been proposed over matrix groups and exhibit their security weaknesses. Once again these proposals are in the two-party scenario and only passive adversaries will be taken into account.

Example 9.2 [Baumslag et al. key transport]
This simple scheme assumes that the two participants, Alice and Bob, share some long-term secret key which specifies two commuting subgroups A, B of a public non-Abelian finitely presented group G. During each execution Bob chooses a session key, and transmits it to Alice in a *three-pass* fashion, as depicted in Figure 9.9.

S. Blackburn, C. Cid, and C. Mullan pointed out a strategy that is likely to succeed for any implementation inspired in the described procedure, when the group G is the group of invertible 4×4 matrices of determinant 1 over the integers $SL_4(\mathbb{Z})$. It was for this case suggested that the commuting subgroups

Alice	Bob
$A, A' \in \mathcal{A}$	$K \in G$
	$B, B' \in \mathcal{B}$

$$\xleftarrow{C = BKB'}$$

$$\xrightarrow{ACA'}$$

$$\xleftarrow{E = AKA'}$$

$$K_{AB} = A^{-1}EA'^{-1}$$

FIGURE 9.9: Baumslag et al. key transport.

$\mathcal{A}$ and $\mathcal{B}$ should be constructed by means of a secret matrix $M \in \mathrm{SL}_4(\mathbb{Z})$, only known to Alice and Bob, so that

$$\mathcal{A} = M^{-1} \begin{pmatrix} \mathrm{SL}_2(\mathbb{Z}) & 0 \\ 0 & I_2 \end{pmatrix} M \text{ and } \mathcal{B} = M^{-1} \begin{pmatrix} I_2 & 0 \\ 0 & \mathrm{SL}_2(\mathbb{Z}) \end{pmatrix} M,$$

where I_2 represents the 2×2 identity matrix. Blackburn et al.'s cryptanalysis builds on the following observations:

i. It suffices to recover K (mod p) for a small number of primes p.

ii. Fixed a prime p, note that there are many equivalent long-term keys, i. e., the matrix M can be replaced by any invertible matrix N for which

$$\mathcal{A} = N^{-1} \begin{pmatrix} \mathrm{SL}_2(\mathbb{Z}) & 0 \\ 0 & I_2 \end{pmatrix} N \text{ and } \mathcal{B} = N^{-1} \begin{pmatrix} I_2 & 0 \\ 0 & \mathrm{SL}_2(\mathbb{Z}) \end{pmatrix} N,$$

namely, any N so that MN^{-1} centralizes both matrices defining the groups $\mathcal{A}$ and $\mathcal{B}$ above.

iii. For fixed p, finding a suitable N as above reduces to solving a rather small set of quadratic equations, which can be done computing a corresponding Gröbner basis.

▯

Example 9.3 [Álvarez et al. key exchange]
Another matrix-based key exchange protocol was proposed by Álvarez et al. in [ATVZ09], where the platform group consists of the 2×2 block upper triangular invertible matrices over a finite field. Essentially, two *high-order* public matrices M_1 and M_2 are generated in this group (the authors of the proposal suggest using companion matrices of primitive polynomials in blocks $(1,1)$ and $(2,2)$ to maximize the order). Then the two users choose secret exponents (r, s) and (v, w), respectively, and exchange the matrices $M_1^r M_2^s$

and $M_1^v M_2^w$ as described in Figure 9.10. The shared key is the $(1,2)$ block of the matrix $M_1^{r+v} M_2^{s+w}$. This is done mainly in order to avoid a reduction from the discrete logarithm problem in the matrix group to the discrete logarithm problem in the base field.

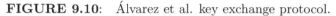

Alice		Bob
r, s		v, w
	$C = M_1^r M_2^s$	
	$\xrightarrow{\hspace{4cm}}$	
	$D = M_1^v M_2^w$	
	$\xleftarrow{\hspace{4cm}}$	
	$K_{AB} = (M_1^r D M_2^s)_{(1,2)} = (M_1^v C M_2^w)_{(1,2)}$	

FIGURE 9.10:   Álvarez et al. key exchange protocol.

It soon became clear that simple linear algebra strategies render the above scheme insecure, as was evidenced in the works [KY13, GVPdPTDV14]. Succintly, the attack from [KY13] states that if one is able to find two matrices W_1 and W_2 commuting, respectively, with M_1 and M_2 and so that $D = W_1 W_2$, then the exchanged key can be computed as $W_1 C W_2$. In other words, the attacker simply faces the following linear system of equations over $\mathbb{Z}_p$.

$$W_1^{-1} M_1 = M_1 W_1^{-1}$$
$$W_2 M_2 = M_2 W_2$$
$$W_1^{-1} D = W_2.$$

⧠

9.5   Summary and further reading

In this chapter we have

- introduced the basic formal definitions related to key exchange protocols and an example of a protocol that can be proven secure within a formal model.

- reviewed a few examples of group-theoretic constructions aiming at secure key exchange in the two-party setting. In particular, we have described the *commutator KEP* proposed by Anshel et al. and the *braid*

Diffie–Hellman protocol of Ko et al. and explained a cryptanalysis technique, based on linear centralizers, that succeeds against these two (and other) constructions. Also, we have described two matrix based protocols which succumb to simple linear algebra attacks.

Key establishment protocols are basic cryptographic constructions widely deployed and frequently integrated as building blocks of more complex constructions. A good book to get into the subject is that of Boyd and Mathuria [BM04]. The Burmester–Desmedt conference key agreement from [BD95] was proven secure in the passive model in [BD05], a variant which provides secure guarantees even if the adversaries are active was later proposed by Katz and Yung [KY07]. Some up-to-date papers refining group key establishment security models are [BCPQ01, BGVS07].

Besides the ones mentioned in this chapter, a number of different proposals have been made for deriving key exchange protocols using non-Abelian groups as a base, including [SU05, Sti05, SU06]. Many of them are only conceptual, and still in most cases the heuristic and linearization attacks described in this chapter evidence they cannot be considered secure [Shp08, Mat08]. A generic framework for building group key exchange protocols from group-theoretic problems can be found in [BGS06], where the approach is to build a so-called *key encapsulation mechanism* from a group-theoretic problem and then construct a group key establishment protocol on top of it. This approach results in a 2-Round construction providing strong guarantees (not only security, in the sense of Definition 9.3, but further properties like *forward secrecy* and *strong entity authentication*). This construction is however conceptual, in the sense that it comes without a proposal of a non-Abelian assumption on which the proposed framework may be materialized.

9.6   Exercises

Exercise 64 *Prove that in the key agreement of Example 9.1 participants are able to establish a common key if only passive adversaries are allowed. Namely, if a suitable identification tag for the key was constructed, the scheme will be correct in the sense of Definition 9.1.*

Exercise 65 *Does the key agreement of Example 9.1 provide* forward secrecy? *Namely, if both* SecSet_i *and* SharSet_i *are disclosed* $(i = 1, \ldots, m)$, *can session keys that have been agreed upon in the past still be considered secure?*

Exercise 66 *Let $\hat{a}$ be an invertible solution of the linear system of equations underlying the braid Diffie–Hellman problem, namely*

$$xg^a = gx$$
$$xb_1 = b_1 x$$
$$\vdots$$
$$xb_k = b_k x.$$

Prove that $\hat{a}$ suffices to compute the element g^{ab}.

Exercise 67 *Consider the protocol from Figure 9.4. Check that it is correct. Analyze the impact on the security of the scheme against active adversaries if the second and third step of the* **Check** *part is eliminated from* **Round 2**; *namely, user U_1 should no longer check*

(a) $T_1 \oplus \cdots \oplus T_n = 0$,

(b) $\mathsf{sid}_i^{s_i} = \mathsf{sid}_j^{s_j}$ $(j = 1, \ldots, n)$,

(c) *the validity of the commitment on k_n.*

Think of the different scenarios that arise if only one of the above checks ((a), (b) or (c)) is suppressed.

Exercise 68 *Think of a different attack strategy for the scheme of Álvarez et al., assuming the matrices M_1 and M_2 can be diagonalized.*

Chapter 10

Signature and identification schemes

10.1 Definitions and terminology

Authenticating an entity or group of entities that "own," have produced, or modified a certain piece of information is central in nearly any scenario where cryptographic techniques are to be applied. As we have seen in the middleperson attack in the previous chapter, even a simple two-party key exchange can be compromised if Alice cannot be assured that she is actually communicating with Bob (see Figure 9.1). Digital signature schemes are the basic tool providing the functionality to solve such a problem in the public-key setting. Given a message m that we may receive, a digital signature on m can evidence who created m, certify it has not been altered, or even be used to confute a malicious sender who denies having sent m.

DEFINITION 10.1 [Public-key signature scheme]   *A public-key signature scheme is given by a triple of polynomial time algorithms $(\mathcal{K}, \mathcal{S}, \mathcal{V})$ along with a family $\{M_\ell\}_{\ell \in \mathbb{N}}$ of message spaces. More specifically, we have the following:*

$\mathcal{K}$: *a probabilistic key generation algorithm that on input a security parameter $1^\ell = \underbrace{1 \ldots 1}_{\ell \text{ times}}$ outputs a pair (pk, sk). We refer to pk as a public key and to sk as a secret key.*

$\mathcal{S}$: *a probabilistic algorithm that on input a secret key sk and a message $m \in M_\ell$ outputs a signature σ for m.*

$\mathcal{V}$: *a deterministic algorithm that on input of a public key pk, a candidate signature σ, and a message $m \in M_\ell$ outputs 1 if the verification is correct (namely, σ is a valid signature for m with the corresponding public key pk) and 0 otherwise.*

For $\mathcal{S}$ and $\mathcal{V}$ we will write the first argument as an index, i.e., we set $\mathcal{S}_{sk}(m) = \mathcal{S}(sk, m)$ and $\mathcal{V}_{pk}(\sigma, m) = \mathcal{V}(pk, \sigma, m)$. Moreover, for key pairs (pk, sk) generated by the key generation algorithm $\mathcal{K}$, we require that the event

$$\exists m \in M_\ell : \mathcal{V}_{pk}(\mathcal{S}_{sk}(m), m) \neq 1$$

occurs with negligible probability only.[1]

As we did in the earlier discussion of public-key encryption, when aiming at formal proofs of security for signature schemes, we have to specify both the goals and capabilities of the adversary. Typically, the adversarial goals are classified as:

- *total break*: disclosing a valid secret key for impersonating the signer at will;

- *universal forgery*: constructing a signing algorithm that on input any message outputs a signature for it that will be accepted as valid;

- *existential forgery*: producing a valid signature of a message m of the adversary's choosing. If m is chosen by the adversary prior to the attack, we speak of *selective forgery*.

Among the above, the more "humble" goal is that of an existential forgery, as it can well be the case that all messages the adversary is able to sign have a particular structure—and are potentially meaningless in the targeted application context.

When it comes to specifying adversarial capabilities there is a first distinction we have to make. Should we assume that the adversary has access to all public information but not to any signed messages (*no-message attack*), or is the adversary granted access to some valid message-signature pairs (*known-message attack*)? Going one step further, should we assume that the adversary has a signing oracle enabling him to sign a large class of messages of his choice? The latter sounds quite similar to the CCA scenario for public-key encryption, and different flavors of such a security notion may be introduced. We include here the (arguably) most used one:

DEFINITION 10.2 [Existential unforgeability]   *For a public-key signature scheme $(\mathcal{K}, \mathcal{S}, \mathcal{V})$ and an algorithm $\mathcal{A}$ consider the experiment in Figure 10.1. We refer to $(\mathcal{K}, \mathcal{S}, \mathcal{V})$ as secure in the sense of* existential unforgeability under chosen message attacks (EUF-CMA) *if for every probabilistic polynomial time algorithm $\mathcal{A}$ the advantage*

$$\mathbf{Adv}_{\mathcal{A},(\mathcal{K},\mathcal{S},\mathcal{V})}(\ell) = \Pr[\mathbf{Exp}_{\mathcal{A},(\mathcal{K},\mathcal{S},\mathcal{V})} = 1] \qquad (10.1)$$

is negligible, even if $\mathcal{A}$ has access to a signing oracle $\mathcal{O}_{\mathcal{S}}$ running the signing algorithm $\mathcal{S}_{sk}(\cdot)$ with the sole restriction that the output message m must not have been queried to $\mathcal{O}_{\mathcal{S}}$.

---

[1]Note again that, as is often the case, probability is considered as a function of the security parameter ℓ.

$$
\begin{array}{|l|}
\hline
\textbf{Exp}_{\mathcal{A},(\mathcal{K},\mathcal{S},\mathcal{V})}\text{EUF-CMA}(1^{\ell}) \\
(pk, sk) \leftarrow \mathcal{K}(1^{\ell}) \\
(m, \sigma) \leftarrow \mathcal{A}^{\mathcal{O}s}(1^{\ell}, pk) \\
\text{if } \mathcal{V}(m, \sigma) = 1 \\
\quad \text{then return } 1 \\
\quad \text{else return } 0 \\
\hline
\end{array}
$$

FIGURE 10.1: Existential unforgeability under chosen message attack.

In an actual use case, one may need more than evidencing that a message has been at a certain point produced by a user and not modified by someone else afterwards. In some cases, we need to be sure that throughout the execution of a process the identity of our partners is the one we presume beforehand. Identification schemes are cryptographic protocols geared towards this goal. They can be classified from what they actually use to identify an entity or individual (see [MvOV96]):

- what they *know*: a cryptographic key is linked inseparably to an identity, this is both the case for public-key identification schemes and secret-key challenge-response protocols;

- what they *have*: entities are identified with a device or token that might itself produce evidence, e. g., a smartcard, a password generator, or a PUF [Pap01];

- what they *are*: physical characteristics or involuntary actions (like handwritten signatures or gait) are used for identifying individuals. *Biometrics* deals with these methods.

We define public-key identification schemes only informally, as we will only provide some examples and avoid the formalization of security notions (which is better introduced in the context of *proofs of knowledge*, such as *zero-knowledge proofs*). Proofs of knowledge are (typically interactive) protocols in which a certain user gives irrefutable assurance of his knowledge of a "critical" piece of information to another user. It is often desirable that this interaction reveals as little as possible about the secret information whose knowledge is being assessed, ideally none at all (this is the case in *zero-knowledge proofs*). A detailed discussion on these topics can be found in [Gol01, Chapter 4].

In a public-key identification scheme two entities are involved: a prover P and a verifier V. The prover generates a public key/private key pair (pk, sk) and uses it to provide evidence of his identity, executing a protocol with the verifier (who holds pk). Security is specified in terms of a passive eavesdropper who may monitor multiple executions of the protocol, or an active one, who may act in the role of the verifier and deviate from the specification of the protocol. Once this interaction phase is over, the goal of the adversary (passive or active) is to successfully impersonate the legitimate prover in a new execution with an honest verifier. Signature schemes trivially yield identification schemes, as the identity of the signer may be proved by simply requesting him to sign randomly generated messages.

10.2 RSA signatures: FDH and PSS

In Example 3.2, we have seen a first example of an RSA-based public-key signature scheme. As already noted there, this straightforward construction is insecure, for given a valid signature σ and a public key (N, e) the adversary can immediately create a valid message/signature pair $(\sigma^e \pmod{N}, \sigma)$. In the random oracle model, a (in the sense of EUF-CMA) secure variant can be constructed easily using the so-called *full domain hash (FDH) paradigm*, introduced by Bellare and Rogaway in [BR96]. This paradigm is summarized in Figure 10.2, where the main tools for signing ℓ-bit messages are a trapdoor one-way permutation defined on a set X of size superpolynomial in ℓ, and a hash function $\mathcal{H}$ with range X and domain $\{0,1\}^\ell$.

$\mathcal{K}$: on input 1^ℓ outputs a description of a trapdoor one-way permutation f defined over X as public key pk, and trapdoor information for evaluating the inverse f^{-1} of f as secret key sk.

$\mathcal{S}$: on input (sk, m), where $m \in \{0,1\}^\ell$, sets $r = \mathcal{H}(m)$, computes $\sigma = f^{-1}(r)$ and returns the signature σ for m.

$\mathcal{V}$: on input (pk, m, σ), checks whether $f(\sigma) = \mathcal{H}(m)$ and returns 1 if this is the case, otherwise it returns 0.

FIGURE 10.2: FDH signature with trapdoor one-way permutation.

THEOREM 10.1 [Security of FDH signature scheme]

The public-key signature scheme depicted in Figure 10.2 is secure in the sense of EUF-CMA in the random oracle model, under the assumption that the function f is chosen from a trapdoor one-way permutation family defined over X.

This security proof is taken from [Poi05], we refer for further details to the original text.

PROOF Let $\mathcal{A}$ be a probabilistic polynomial time adversary that is able to produce, with non-negligible probability ε, an existential forgery under a chosen message attack for the scheme depicted in Figure 10.2. That is, on input the public key and having access to a signing oracle $\mathcal{O}_S$ outputs a message/signature pair that will be accepted by $\mathcal{V}$ on input pk, where the output message m has never been queried to $\mathcal{O}_S$. More precisely we have

[Signing oracle $\mathcal{O}_S$]: on input a message m it

1. sets $r = \mathcal{H}(m)$,

2. computes $\sigma = f^{-1}(r)$, using the secret key sk,

3. outputs (m, σ).

As we did earlier, we use the game-hopping technique. The adversarial probability of success for Game i will be denoted by P_i.

Game G_0: This is exactly the "real" game modeling $\mathcal{A}$'s attack, thus, we have to prove that its success probability (denoted by P_0) is negligible in the security parameter. To do so, choose $y \leftarrow X$ uniformly at random. We will modify $\mathcal{A}$ such that it will produce a preimage $f^{-1}(y)$ of y.

Game G_1: At this, the random oracle $\mathcal{H}$ is simulated in perfect agreement with the random oracle model, namely:

[$\mathcal{H}$-oracle]: on input a new value m, it selects $r \in X$ uniformly at random and stores (r, m) on an (indexed) list as $(r, \perp, m)$—the field $\perp$ is meaningless at this point, but will play a role later. If $\mathcal{H}$ is then queried again on m, the output will be r.

Accordingly, the use of the random oracle $\mathcal{H}$ is replaced by this simulated $\mathcal{H}$-oracle in the description of the signing oracle $\mathcal{O}_S$ and also in the verification step, which we impose on the adversary without loss of generality: Before producing an output (m, σ) as a claimed forgery, we impose that the adversary queries the $\mathcal{H}$-oracle on the message m and then checks whether its output matches $f(\sigma)$. As the simulation of the random oracle is perfect, we have $P_1 = P_0$.

Game G_2: In this game the challenger guesses (a priori) uniformly at random the index i_c when the adversary submits the message m for the forgery to the $\mathcal{H}$-oracle. Whenever this guess is wrong, the adversary is considered at a loss. With probability at least $1/(q_h + q_s + 1)$ the guess for i_c is correct, where the denominator is a polynomial upper bound for the total number of queries to the $\mathcal{H}$-oracle (q_h denotes a polynomial upper bound for number of direct queries to $\mathcal{H}$, q_s a polynomial upper bound for the number of queries to the signing oracle $\mathcal{O}_S$, and the 1 comes from the $\mathcal{H}$-query at the verification step). Thus,

$$P_2 \geq \frac{P_1}{q_h + q_s + 1}.$$

Game G_3: Now the simulation of the oracle $\mathcal{H}$ is modified. In exactly the query corresponding to m, instead of selecting $r \in X$ uniformly at random, use as $\mathcal{H}$-image the challenge value y for which $f^{-1}(y)$ has to be found. Indeed, as the challenge was selected also uniformly at random from X, we have $P_3 = P_2$.

> **[$\mathcal{H}$-oracle]:** For query i_c output y and insert $(y, \perp, m)$ into the list. Otherwise, select $r \in X$ uniformly at random and store $(r, \perp, m)$ as before.

Game G_4: We now modify the oracle $\mathcal{H}$ for all but the exact query i_c. Namely,

> **[$\mathcal{H}$-oracle]:** For query i_c, as before. For any other query q, select $s \in X$ uniformly at random and set $r = f(s)$. Store (r, s, q) in the list.

Because of f being a permutation we have $P_3 = P_4$.

Game G_5: Let us now change the signing oracle so that it never evaluates f^{-1}, but simply reads off from the list created by the $\mathcal{H}$-simulation. On input any query q to the signing oracle, we may simply look for the entry (r, s, q) and output $\sigma = s$. As this modification does not affect the success of the forgery (for the message corresponding to the i_c query cannot be queried to the signing oracle), we have $P_5 = P_4$.

Clearly, P_5 is exactly the probability the adversary has of successfully computing the preimage $f^{-1}(y)$ of y (and no oracles are needed to find this preimage). Due to the one-wayness of f, the latter probability is negligible. This concludes the proof.

□

The above proof actually gives a precise reduction, namely, if ε is the success probability of the adversary against the signature scheme, he will invert the permutation f with probability at least $\frac{\varepsilon}{q_s + q_h + 1}$. Actually, when looking more closely at exact reductions, it makes a difference if the one-way permutation f is homomorphic—with respect to a product defined in X—as is the case for RSA.

Aiming at a tighter reduction, M. Bellare and P. Rogaway also introduced in [BR96] a variant of FDH called the *probabilistic signature scheme (PSS)*. This scheme makes use of three hash functions (idealized in the proof as random oracles). Assuming the elements from X are encoded with k bits, consider three hash functions

$$\mathcal{F} : \{0,1\}^{k_2} \longrightarrow \{0,1\}^{k_0}$$
$$\mathcal{G} : \{0,1\}^{k_2} \longrightarrow \{0,1\}^{k_1}$$
$$\mathcal{H} : \{0,1\}^* \longrightarrow \{0,1\}^{k_2}$$

where $k = k_0 + k_1 + k_2 + 1$. The PSS signature scheme is described in Figure 10.3.

$\mathcal{K}$: on input 1^ℓ outputs a description of a trapdoor one-way permutation f as public key pk, and trapdoor information for evaluating the inverse f^{-1} of f as secret key sk.

$\mathcal{S}$: on input (sk, m), chooses a random string $r \in \{0,1\}^{k_1}$ and sets

$$w = \mathcal{H}(m, r), s = \mathcal{G}(w) \oplus r \text{ and } t = \mathcal{F}(w).$$

Further, set $y = 0||w||s||t$, $\sigma = f^{-1}(y)$, and output the signature σ.

$\mathcal{V}$: on input (pk, m, σ),

1. compute $y = f(\sigma)$ and parse it as $a||b||c||d$ with $(a, b, c, d) \in \{0,1\} \times \{0,1\}^{k_2} \times \{0,1\}^{k_1} \times \{0,1\}^{k_0}$,
2. return 1 if $a = 0$, $b = \mathcal{H}(m, c \oplus \mathcal{G}(b))$, and $t = \mathcal{F}(w)$, otherwise return 0.

FIGURE 10.3: PSS signature with trapdoor one-way permutation.

Both FDH and PSS have been extensively deployed using the RSA (candidate) trapdoor one-way permutation, and thus are typically regarded as RSA signatures. Having in mind non-Abelian groups, logarithmic signatures as used in the MST_1 public-key encryption scheme by Magliveras et al. [MSvT02] come to mind—if cryptographically acceptable parameters can be identified, then such logarithmic signatures provide a candidate for a trapdoor one-way permutation on $\mathbb{Z}_{|G|}$ for a finite group G. If one is tempted to follow the above construction to derive a signature scheme based on the hardness of the discrete logarithm problem, one encounters the problem that a trapdoor is needed, which in commonly used cyclic groups is not available. So far, the Digital Signature Algorithm DSA is perhaps the most prominent public-key signature scheme based on a discrete logarithm problem (see Figure 10.4). It was adopted by NIST as a standard in 1994 (see [NIS94]). An efficient variant is ECDSA, which is defined over a group structure from certain elliptic curves.

$\mathcal{K}$: on input 1^ℓ selects an ℓ-bit prime p and a prime q dividing $p-1$. Further, selects $g \neq 1$ so that $g^q = 1 \pmod{p}$. Finally, choose $x \leftarrow \mathbb{Z}_q$ and set $y = g^x \pmod{p}$.

The output keys are then defined as $pk = (p, q, g, y)$ and $sk = x$.

$\mathcal{S}$: On input the secret key sk, choose $k \leftarrow \mathbb{Z}_q^*$, set

$$r = (g^k \bmod p) \bmod q \text{ and}$$
$$s = k^{-1}(H(m) + xr) \bmod q.$$

Output the signature $\sigma = (s, r)$.

$\mathcal{V}$: On input a public key $pk = (p, q, g, y)$ and a message/signature pair $(m, (s, r))$,

1. check if $r, s \in \{1, \ldots, q-1\}$, otherwise output 0,
2. set

$$w = s^{-1} \bmod q,$$
$$u_1 = H(m)w \bmod q,$$
$$u_2 = rw \bmod q,$$

3. compute $v = (g^{u_1} y^{u_2} \bmod p) \bmod q$,
4. output 1 if $v = r$, and 0 otherwise.

FIGURE 10.4: Digital Signature Algorithm.

10.3   Identification schemes

Identification schemes can be seen as special *proofs of knowledge*; an entity is identified by proving (typically, disclosing as little information as possible) knowledge of critical data only this entity is supposed to hold. A general technique (due to U. Feige, A. Fiat, and A. Shamir [FFS88]) to derive identification schemes is based on one-way functions and essentially builds on the idea that a prover, holding a randomly selected value r, may publish $y = f(r)$ for a public one-way function f and later identify himself by proving knowledge of an f-preimage for y. The *proof of knowledge* should not disclose any information useful to subsequently compute f-inverses. It can actually be formally proved that secure identification is possible if one-way functions exist.

Let us informally describe the so-called Fiat–Shamir identification scheme, which is one of the most prominent examples of the above generic construction. Once a security parameter ℓ is fixed, the prover selects uniformly at random N, a product of two ℓ-bit long prime numbers, and $s \in \mathbb{Z}_N^*$. The prover further sets

$$t = s^2 \pmod{N}$$

and publishes (N, t). To prove his identity, he must convince the verifier of his knowledge of a square root of t, which he achieves by repeating the steps depicted in Figure 10.5 sufficiently many times.

Prover	Verifier
$g \leftarrow \mathbb{Z}_N, h = g^2 \pmod{N}$	
$\xrightarrow{\quad h \quad}$	
	$\sigma \leftarrow \{0, 1\}$
$\xleftarrow{\quad \sigma \quad}$	
$a = g s^\sigma \pmod{N}$	
$\xrightarrow{\quad a \quad}$	
	accepts iff
	$a^2 = h t^\sigma \pmod{N}$

FIGURE 10.5: Fiat–Shamir identification.

In order to evaluate an identification scheme, there are three properties that must be carefully analyzed. We describe them informally:

- *Completeness*: if both participants follow the protocol specification, the verifier will accept the input provided by the prover with overwhelming probability;

- *Soundness*: anyone who succeeds in impersonating the prover is also able to compute his private key from the public information;

- *Zero-Knowledge*: the verifier learns nothing beyond the validity of the proof.

Certainly, completeness is a strict requirement, whereas the soundness and zero-knowledge properties can be graded or loosened up depending on the application scenario. For the case of the basic version of Fiat–Shamir identification in Figure 10.5, completeness is straightforward to verify, while soundness comes with a probabilistic argument. Namely, if the identification step is repeated k times, the probability of an adversary successfully impersonating the legitimate prover is bounded by $2^{-k} + P_{sqrt}$, where P_{sqrt} is the probability he has of computing a square root of $t \bmod N$ from the public information.

The literature contains a few interesting attempts to adapt the Fiat–Shamir paradigm to a non-Abelian setting. We describe here two braid-based proposals, which follow the lines of the schemes we already discussed in Chapter 6 and Chapter 9. Both can be found in the paper [SDG06].

Example 10.1 [Identification from braid groups]
Based on a security parameter ℓ, a braid group B_n for some n and a hash function

$$H : B_n \longrightarrow \{0,1\}^k,$$

for k polynomial in ℓ are made publicly available to all users. The prover selects $b \leftarrow B_n$ and $s \leftarrow LB_n$ (a random distribution needs to be chosen here), sets

$$b' = sbs^{-1}$$

and publishes (b, b'). To identify himself to the verifier, two different identification phases have been proposed, one based on the braid Diffie–Hellman search problem (see Figure 10.6) and another one relying on the hardness of the conjugacy search problem (described in Figure 10.7). The latter identification phase is assumed to be repeated a number of times, in order to provide a certain level of soundness.

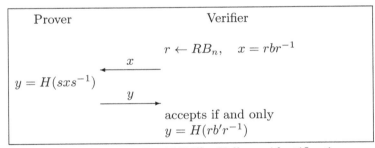

FIGURE 10.6: Braid Diffie–Hellman identification.

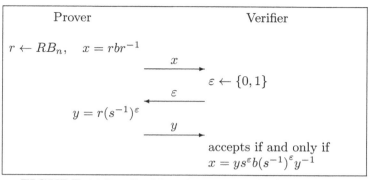

FIGURE 10.7:   Braid conjugacy search identification.

10.4   Summary and further reading

In this chapter we have

- defined public-key signature scheme, and discussed how to prove them secure in the sense of EUF-CMA. As prominent examples we have seen RSA-based constructions and DSA;

- introduced the notion of identification scheme and briefly introduced the generic Fiat–Shamir construction;

- informally described two proposals using braid groups to construct an identification scheme.

There are many different application scenarios where the need for a dedicated signature procedure arises. Such applications are behind the introduction of group signatures, where any member of a specific group is authorized to sign on behalf of the group, maintaining a certain level of anonymity as individual. Ring signatures are a special type of group signatures, with the additional property that any subgroup of users can at a certain point constitute a new signing group without any additional setup. Also, the anonymity of users is, unlike in group signatures, unrevokable. When designing signature schemes it may also be interesting to impose threshold conditions both at the signing and at the verification phase; if we go for threshold signing, a minimum of t out of n users from a group are needed in order to form a valid signature, while any smaller coalition is not supposed to create a valid signature or prevent the legitimate group from doing so (see, for instance [Sho00]). On the other hand, if the threshold phase is at the verification, we wish to enforce that no verifier can gain advantage from stating the authenticity of a message before the other group members assess its validity [PM97]. Numerous other examples of application-tailored signature schemes can be found in the literature, including undeniable, fail-stop, one-time signatures, etc. A recent monograph on the subject is [Kat10].

Identification schemes are a subclass of so-called *proof of knowledge* protocols, which are very powerful tools within modern cryptology. Zero-knowledge proofs of knowledge play a central role in the design of commitment schemes and are behind the IND-CCA security of many public-key encryption schemes. Indeed, to render the decryption oracle in the IND-CCA scenario useless, it suffices to require that a valid ciphertext should always include a zero-knowledge proof of knowledge of the plaintext. For more information on this topic we refer the interested reader to [Gol01, Chapter 4]. For recent results on the connections between signature and identification schemes, see [KH06].

10.5 Exercises

Exercise 69 *Check the correctness of the DSA algorithm, namely, verify that if the scheme specification is followed faithfully, then a signature pair will be recognized as valid by the verification algorithm with overwhelming probability.*

Exercise 70 *Argue how security is compromised if the same value k in the DSA signature algorithm is used for signing two different messages.*

Exercise 71 *Design an identification scheme from a public-key encryption that is secure in the sense of IND-CCA. Give a bound on the probability of success the adversary has in impersonating the legitimate prover.*

Exercise 72 *Consider the identification scheme plainly derived from a public-key signature scheme. Namely, the verifier, holding the public key of the signature scheme, challenges the prover with n randomly selected messages from which he should derive valid signatures. Give a bound on the probability of success a passive adversary has in impersonating the legitimate prover, provided that you have a bound for his probability of successfully attacking the signature scheme in a certain sense.*

Exercise 73 *One-time signatures are, informally, signature schemes whose security properties are only guaranteed insofar as a single message is signed using each output of the key generation algorithm. Consider the signature scheme from Figure 10.8, defined for messages of length l (where l is actually a function of the security parameter ℓ). Let $\mathcal{A}$ be a successful forger achieving with non-negligible probability a valid message/signature pair (m, σ) with the sole input being the public key pk and granted a single call to a signing oracle* *that he may use once having pk. Prove that $\mathcal{A}$ can be used to derive an algorithm to invert f.*

$\mathcal{K}$: on input 1^{ℓ} output a description of a one-way function f with domain $\{0, 1\}^{\ell}$ and

    1. select $x_{0,i}, x_{1,i} \leftarrow \{0, 1\}^{\ell}$ for $i = 1, \ldots l$

    2. compute $y_{0,i} = f(x_{0,i})$ and $y_{1,i} = f(x_{1,i})$ for $i = 1, \ldots, l$

    3. output (pk, sk), where

$$pk = \begin{pmatrix} y_{0,1} & y_{0,2} & \cdots & y_{0,l} \\ y_{1,1} & y_{1,2} & \cdots & y_{1,l} \end{pmatrix} \text{ and } sk = \begin{pmatrix} x_{0,1} & x_{0,2} & \cdots & x_{0,l} \\ x_{1,1} & x_{1,2} & \cdots & x_{1,l} \end{pmatrix}.$$

$\mathcal{S}$: on input sk as above and $m = m_1 || \ldots || m_l \in \{0, 1\}^l$ set $\sigma = (x_{m_1,1}m, \ldots, x_{m_l,l})$ and output σ.

$\mathcal{V}$: on input pk as above, $m = m_1 || \ldots || m_l \in \{0, 1\}^l$, and a candidate signature $\sigma = \sigma_1 || \ldots || \sigma_l$, output 1 if $f(\sigma_i) = y_{m_i,i}$ for $i = 1, \ldots, l$, and return 0 otherwise.

FIGURE 10.8: Lamport's one-time signature.

Exercise 74 *When using signatures as part of the session identifier in a key establishment protocol (cf. Chapter 9), it can be of interest to impose strong unforgeability on a signature scheme. The corresponding definition is identical to the definition of EUF-CMA with one difference: The adversary succeeds if he produces a message/signature pair (m, σ) such that σ has not been obtained from the signing oracle. In particular, obtaining a new signature on an already signed message is considered a successful attack.*

Does the FDH signature scheme meet this stronger unforgeability requirement?

Exercise 75 *Analyze the soundness of the two identification protocols from Example 10.1.*

Part V

Appendix

Appendix A

Solutions to selected exercises

A.1 Solutions to selected exercises of Part I

Solution to Exercise 2. It is clear that f_g is bijective. To verify that f_g is a homomorphism, take $h, k \in G$. Then $f_g(hk) = g^{-1}hkg = g^{-1}hgg^{-1}kg = f_g(h)f_g(k)$, which establishes (a). For (b), a direct computation shows that $f_g \circ f_h = f_{hg}$ and $f_g^{-1} = f_{g^{-1}}$. As a result, $\mathrm{Inn}(g) \leq \mathrm{Sym}(G)$ and $\mathrm{Inn}(g) \leq \mathrm{Aut}(G)$. Finally, to establish (c) we have to see that conjugates of elements from $\mathrm{Inn}(G)$ are also in $\mathrm{Inn}(G)$. Indeed, let $\Psi \in \mathrm{Aut}(G)$ be any automorphism of G. Now for any $x \in G$ we have $(\Psi^{-1}f_g\Psi)(x) = (\Psi^{-1}f_g)(\Psi(x)) = \Psi^{-1}(g^{-1}\Psi(x)g) = \Psi^{-1}(g^{-1}) \cdot x \cdot \Psi^{-1}(g) = f_{\Psi^{-1}(g)}(x)$.

Solution to Exercise 4. Let $g \in G$ with $g \neq 1$. As the order of g must divide $|G| = p$, the order of g has to be p. As a consequence, $G = <g>$, i.e., g generates the full group.

Solution to Exercise 6. This follows from the fact that Sylow p-subgroups are conjugate. Clearly, H must be contained in a Sylow p-subgroup P (or be itself a Sylow p-subgroup P). Because of H being normal, then it will be contained in all.

Solution to Exercise 8. Just apply Lagrange's theorem to $\mathbb{Z}_n^*$.

Solution to Exercise 10. In the quaternion group, the three subgroups of order four are all cyclic. This is not the case for D_4 which also has three subgroups of order 4, but one of them is cyclic and the others are isomorphic to $\mathbb{Z}_2 \times \mathbb{Z}_2$. A shorter argument follows by counting elements of order 2.

Solution to Exercise 12. It is enough to argue that for any polynomial $p(n)$ there exists a constant $c \geq 0$ such that $p(n) \leq n^c$ for every natural number n.

Solution to Exercise 14. It is easy to see that the algorithm is quadratic if one compares it with the Euclidean algorithm from Example 2.3. Indeed, the while loop in subroutine 1 executes as many divisions – starting from $a = mq + r$ — as it would do for computing $gcd(a, m)$, as a result, its running time is also in $\mathcal{O}(\ln^2 m)$.

Solution to Exercise 16. This problem is a simple version of the (much more challenging) general *ideal membership problem* in polynomial rings with $r \geq 1$ indeterminates.

To test if $g \in I$, it suffices to check whether $\gcd(f_1, \ldots, f_n)$ divides g. This greatest common divisor may be computed from an analogue of the Euclidean algorithm for the case of natural numbers. Instead of ordinary integer division, we now use long division of polynomials.

Solution to Exercise 18. First of all you have to decide what is the block size of the encryption method, i.e., the value n. As the text is of length 12, it can either be $2, 3, 6$, or 12. Let us start with 2. Next, construct a system linking ciphertext and plaintext substrings as follows:

$$E \begin{pmatrix} 13 & 11 \\ 3 & 18 \end{pmatrix} = \begin{pmatrix} 15 & 10 \\ 25 & 17 \end{pmatrix}.$$

When selecting this, we have tried to construct a system with a unique solution, i.e., we selected the plaintext matrix

$$E \begin{pmatrix} 13 & 11 \\ 3 & 18 \end{pmatrix}$$

because it is invertible mod 26. The matrix $E = \begin{pmatrix} 2 & 5 \\ 1 & 4 \end{pmatrix}$ is thus computed as

$$\begin{pmatrix} 15 & 10 \\ 25 & 17 \end{pmatrix} \begin{pmatrix} 13 & 11 \\ 3 & 18 \end{pmatrix}^{-1}$$

and is actually the key that was used to construct the ciphertext.

Solution to Exercise 20. The proof is easily completed; just argue that $(m - \nu k)(m + \nu k)$ must be divided by n.

Solution to Exercise 21. Alice must start by computing s, the multiplicative inverse of r in $\mathbb{Z}_q$ (she may do so using the extended Euclidean algorithm). Now $cs \equiv \sum_{i=1}^{n} b_i \beta_i s \pmod{q}$ and thus $cs \equiv \sum_{i=1}^{n} b_i w_i \pmod{q}$. Note that actually $cs \in \{0, \ldots, q-1\}$. Now the secret message may be retrieved using a greedy strategy; if $w_n > cs$, then $b_n = 0$, otherwise, $b_n = 1$. Now set $cs = cs - w_n b_n$ and do the same with $n-1, n-2$ to 1. The assumption underlying Merkle–Hellman encryption is that the corresponding (so-called *Knapsack*) problem is hard; namely: *given an arbitrary sequence $\beta = (\beta_1, \ldots, \beta_n)$ in $\mathbb{Z}_q$, and a positive integer S, it must be hard to compute a n-bit string $b_0 \ldots b_{n-1}$ such that $S = b_0 \ldots b_{n-1}$.*

Solution to Exercise 22. Verification must be performed via two checks; on input a message m and a signature (r, h, s), the latter will only be accepted as valid if both $h = \mathcal{H}(m, r)$ and $r = g^s y^h$. At least two assumptions we should clearly take; that it is not easy to provide preimages of given h values for the function $\mathcal{H}$ and that discrete logarithms on G for base g are hard to compute.

Solution to Exercise 23. Indeed, suppose Alice may construct a pair $(v', r') \neq (v, r)$ so that $g^{v'} h^{r'} = C = g^v h^r$, then $v' + lr' \equiv s + lr \pmod{q}$ and as a result $l \equiv \frac{v - v'}{r - r'} \bmod q$. Note that, as q is primer and $r \neq r'$, $r - r'$ is a unit in $\mathbb{Z}_q$.

A.2   Solutions to selected exercises of Part II

Solution to Exercise 24. It is easy to see that the security of the proposed two variants of the Bellare–Rogaway construction (Figure 4.8) is by no means as high as that of the original construction:

[BR-α] $\mathcal{E}$, *on input* (pk, m), *where* $m \in \{0, 1\}^\ell$, *selects uniformly at random a value* r *in* X *and computes*

$$x = f(r), \ y = m \oplus G(r) \ and \ z = H(m).$$

It is clear that the resulting scheme is not even IND-CPA: an adversary, getting the encryption corresponding to a message m_b gets as part of the challenge ciphertext the value $H(m_b)$. As he holds m_0 and m_1, he can simply hash and compare for deciding with overwhelming success probability which of the plaintexts corresponds to the challenge ciphertext.

[BR-β] $\mathcal{E}$, *on input* (pk, m), *where* $m \in \{0, 1\}^\ell$, *selects uniformly at random a value* r *in* $\{0, 1\}^\ell$ *and computes*

$$x = f(r), \ y = m \oplus G(r) \ and \ z = H(r).$$

In this case it is easy to see that, even in the absence of a decryption oracle, an adversary can construct new valid ciphertexts from an eavesdropped one. Consequently, the scheme is no longer secure in the sense of NM-CPA. Indeed, just adding (modulo 2) any bitstring $\widetilde{m}$ of the appropriate length to the y-part of the ciphertext leads to a valid encryption of $m \oplus \widetilde{m}$.

Solution to Exercise 26. The key point behind the solution is to take any IND-CPA encryption scheme with a "small" message space. As the success probability of the adversary will always be lower-bounded by $\frac{1}{|\mathcal{M}_\ell|}$, if we have $|\mathcal{M}_\ell| \leq p(\ell)$ for a polynomial p, we get the non-negligible bound.

Solution to Exercise 28. The adversary holds $(g^\beta, g^{\alpha\beta} \cdot m)$. Inverting both components, he simply constructs $(g^{-\beta}, g^{-\alpha\beta} \cdot m^{-1})$ which is a valid encryption for m^{-1}.

Solution to Exercise 30. Actually, such an example can only be found if the family of negligible functions is uncountable; the interested reader can look deeper into this topic in [Bel97]. Let us then take the collection of all negligible functions mapping $\mathbb{N}$ to $\mathbb{R}_{\geq 0}$. It is easy to see that there is no function n fulfilling the required property, as, if n were such function, then $2n$ would be also negligible, but $2n(\ell) \geq n(\ell)$ for all $\ell \in \mathbb{N}$.

Solution to Exercise 32. Take g any generator of G. Then $\alpha = [\alpha_1, \ldots, \alpha_\ell]$ with

$$\alpha_i = [\epsilon_H, g^{p^{i-1}}, \ldots, g^{(p-1)p^{i-1}}]$$

for $i = 1, \ldots, \ell$ is a logarithmic signature for H of length $\ell \cdot p$. Furthermore, as for $i = 1, \ldots, \ell$ and $j = 0, \ldots, p-1$ we have $\alpha_{ij} = g^{jp^{i-1}}$ if for some $h \in G$ we have $h = \alpha_{1i_1} \cdots \alpha_{\ell j_\ell}$, then we have $h = g^k$ with $k = i_1 + i_2 p + \cdots i_\ell p^{\ell-1}$.

Solution to Exercise 34. In the proposed variation of the encryption algorithm in Cramer–Shoup '98, a challenge ciphertext of the IND-CCA game has the form:

$$c_b = (u_1, u_2, h^r m_b, \alpha), \text{ where } \alpha = H(u_1, u_2, h^r).$$

A passive adversary may simply multiply the third component of c_b with the inverse of m_1 and then hash the result together with u_1 and u_2, which should coincide with the received α if the guess $b = 1$ was correct. Trivially, he succeeds in winning the IND-CCA game with non-negligible probability.

Solution to Exercise 36. We are asked to characterize the subgroup P of $\mathbb{Z}_{N^2}^*$ consisting of all N^{th} powers of elements in $\mathbb{Z}_{N^2}^*$, namely, all

$$y \in \mathbb{Z}_{N^2}^* \text{ such that } \exists x \in \mathbb{Z}_{N^2}^* \text{ with } y = x^N \pmod{N^2}.$$

Using the group isomorphism f we gather that

$$y \in P \Rightarrow f^{-1}(y) = (a, b)^N = (Na, b^N) = (0, b^N)$$

where $(a, b) = f^{-1}(x)$ with $x \in \mathbb{Z}_{N^2}^*$ such that $y = x^N \pmod{N^2}$. As a result, we have that any $y \in P$ can be mapped to an element in $\mathbb{Z}_N^*$.

Reciprocally, given any $b \in \mathbb{Z}_N^*$, we may see that $f(0, b) \in P$. Let us start by recalling that, as $\gcd(N, \varphi(N)) = 1$, there exists a $d \in \{0, \ldots, \varphi(N) - 1\}$ such that $Nd \equiv 1 \pmod{\varphi(N)}$. Now, for any $b \in \mathbb{Z}_N^*$, we are going to find an Nth root of $y = f(0, b)$. Indeed, let a be any element in $\mathbb{Z}_N$ and take $x = (1 + N)^a b^d \pmod{N^2}$. Clearly, $f^{-1}(x^N) = (Na, b^{dN}) = (0, b)$.

Thus, P can be seen as the set

$$\{b \mid b \in \mathbb{Z}_N^*\}.$$

The above also proves that P is of size $\varphi(N)$, whereas $\mathbb{Z}_{N^2}^*$ is of size $N\varphi(N)$. Moreover, we have identified all N^{th} roots of any $y \in P$; those corresponding to pairs (a, b) where $a \in \mathbb{Z}_N$ for b such that $f^{-1}(y) = (0, b)$.

Solution to Exercise 38. Let us consider the vector space $X = \mathbb{F}_q^n$, with q prime, and $\{\alpha_1, \ldots, \alpha_n\}$ a $\mathbb{F}_q$-basis of X. Consider H a subgroup of $\mathrm{GL}_n(\mathbb{F}_q)$, leaving a d-dimensional subspace L invariant.

Now, for a given $M \in \mathrm{GL}_n(\mathbb{F}_q)$ denote by M_d the matrix representing the linear transformation induced by M on L. Clearly,

$$\chi : H \longrightarrow \mathrm{GL}_d(\mathbb{F}_q)$$
$$M \longmapsto \quad M_d,$$

is a group homomorphism.

Now, diversity of the corresponding group action system $(X, H, \chi, \mathrm{GL}_d(\mathbb{F}_q))$ can be proven if H is chosen with care (ideally, $|H|$ should only have large prime divisors).

Solution to Exercise 40. It suffices to take $a = \sigma_1 \sigma_2 \sigma_3$ and $b = \sigma_1 \sigma_2$. It is easy to see that $a^2 = b^3$. Further, just note that $\sigma_1 = ba^{-1}$ and as a result σ_1 is also in $\langle a, b \mid a^2 = b^3 \rangle$.

Solution to Exercise 42. For each $v \in V$ we introduce variables $x_{v,\mathrm{b}}$, $x_{v,\mathrm{g}}$, $x_{v,\mathrm{r}}$, and we define the following system of equations over $\mathbb{F}_2$:

$$x_{v,c}^2 + x_{v,c} = 0 \quad (v \in V, c \in \{\mathrm{b}, \mathrm{g}, \mathrm{r}\}) \qquad (\mathrm{A.1})$$
$$x_{v,\mathrm{b}} + x_{v,\mathrm{g}} + x_{v,\mathrm{r}} = 1 \quad (v \in V) \qquad (\mathrm{A.2})$$
$$x_{v,\mathrm{b}} x_{v,\mathrm{g}} + x_{v,\mathrm{b}} x_{v,\mathrm{r}} + x_{v,\mathrm{g}} x_{v,\mathrm{r}} = 0 \qquad (\mathrm{A.3})$$
$$x_{v,\mathrm{b}} x_{v',\mathrm{b}} + x_{v,\mathrm{g}} x_{v',\mathrm{g}} + x_{v,\mathrm{r}} x_{v',\mathrm{r}} = 0 \quad (\{v, v'\} \in E) \qquad (\mathrm{A.4})$$

Equations of type (A.1) ensure that we are considering solutions in $\mathbb{F}_2$ only, equations of type (A.2) ensure that each vertex has a color, and equations of type (A.3) ensure that a vertex has not more than one color. Finally, equations of type (A.4) ensure that there is no edge between vertices of the same color.

Solution to Exercise 44. First one observes that, in cycle notation, conjugating a permutation π to $\tau^{-1} \circ \pi \circ \tau$ with some permutation $\tau \in S_n$ translates into applying τ^{-1} to each element in each cycle of π. So if two given permutations have different *cycle types* they cannot be conjugate—for each cycle length, two conjugate permutations must contain exactly the same number of cycles.

If the cycle types coincide, we can simply define τ^{-1} (respectively τ) by pairing cycles of identical length, mapping the elements in each cycle of one permutation in order to elements in a cycle of the same length in the other permutation.

Solution to Exercise 46. Linear algebra attacks are (too) powerful here. In particular, an algorithm for computing the Frobenius normal form can be used to decide if two elements in $\mathrm{GL}_n(\mathbb{F}_p)$ are conjugate (cf., e g., [Sto98]).

A.3   Solutions to selected exercises of Part III

Solution to Exercise 48. The permutation inv leaves 0 and 1 invariant. Moreover, for $\alpha \in F_{256} \setminus \{0, 1\}$ we have $\alpha \neq \alpha^{-1}$ and $(\alpha^{-1})^{-1} = \alpha$. Hence, inv is the product of $(256 - 2)/2 = 127$ transpositions, and therewith odd.

Solution to Exercise 50. From the key schedule and the definition of f_{K_i}, it follows that

$$\mathrm{DES}_{\overline{K}}(\overline{m}) = \overline{c}$$

with $\bar{\cdot}$ denoting bit-wise complement, i.e., adding an all-one-vector of the appropriate length. So instead of exhausting a space of 2^{56} possible keys of length 56 (there are eight parity bits), we can restrict to enumerating only 2^{55} keys. More specifically, with a plaintext-ciphertext pair (m, c) we can exhaust all keys K_0 where one bit is fixed to 0 and compute $c_0 = \mathrm{DES}_{K_0}(m)$ for each of these keys. If $c_0 = c$, we identify $K = K_0$ as the secret key. If $c_0 = \overline{c}$, we conclude that $\mathrm{DES}_{\overline{K_0}}(\overline{m}) = \overline{c}$ and identify $K = \overline{K_0}$ as the secret key.

Solution to Exercise 52. For Part (a), setting e and g equal to the identity, we see that for any $f \in E$, also f^{-1} is in E. Similarly, choosing f as the identity, we recognize that for any $e, g \in E$ also their product $g \circ e$ is contained in E. Consequently, E is a group.

For part (b), we exploit that A_n is a normal subgroup of S_n. In particular this implies that left- and right cosets of A_n are identical. Let us assume that $e, f, g \in k \circ A_n$. Then we have $e = k \circ a_e$, $f = k \circ a_f$, and $g = k \circ a_g$ for some $a_e, a_f, a_g \in A_n$, and $k \in S_n$. We obtain

$$g \circ f^{-1} \circ e = \underbrace{k \circ a_e \circ a_f^{-1} \circ k^{-1}}_{\in A_n} \circ k \circ a_e \in \underbrace{A_n \circ k}_{= k \circ A_n} \circ a_e,$$

and the claim follows.

Solution to Exercise 54. The error in the first ciphertext block c_1 results in an error in the first plaintext block m_1. Moreover, the "masking value" e_2 is wrong, so the decryption of the second ciphertext block is affected. All other blocks, beginning with c_3, are decrypted correctly.

Solution to Exercise 56. As h is a single non-injective and fixed function, there exist two inputs $m_0 \neq m_1$ which satisfy $h(m_0) = h(m_1)$. The values m_0 and m_1 do not depend on ℓ, so there is a polynomial time algorithm $\mathcal{A}$ which always outputs m_0 and m_1, and this algorithm produces a collision for h with success probability 1.

Solution to Exercise 58. To find a collision it suffices to find a representation (of positive length) of the identity in S_n in terms of the generators S. In particular, if we choose ν as the least common multiple of the cycle lengths occurring in the representation of an element $s \in S$, then ν is the order of s. Consequently, s^ν is the identity, and we can insert ν repetitions of the plaintext symbol corresponding to s into any message without affecting the hash value.

(The largest possible value of ν is bounded by *Landau's function*, which for $n \in \mathbb{N}$ is defined to be the largest order of an element in S_n. The reader interested in learning more about this function may want to take a look into a paper by W. Miller [Mil87].)

Solution to Exercise 60. To find a suitable polynomial $f(x)$, we can first define "generic generators"

$$G_0 = \begin{pmatrix} x & 1 \\ 1 & 0 \end{pmatrix}, G_1 = \begin{pmatrix} x & x+1 \\ 1 & 1 \end{pmatrix} \in \mathrm{SL}_2(\mathbb{F}_2[x])$$

with polynomial entries. Now we can fix some bitstring $b \in \{0,1\}^*$ and, using these generic generators, hash

$$b^\nu = \underbrace{b||\ldots||b}_{\nu \text{ repetitions}}$$

for some small natural number ν. The idea is that b hashes onto an element of order ν when replacing the variable x with an appropriate field element α, so that ν repetitions of b hash to the identity.

Let $H_{b^\nu}(x)$ be the generic hash value of the bitstring b^ν in $\mathrm{SL}_2(\mathbb{F}_2[x])$. We can subtract the identity matrix from $H_{b^\nu}(x)$ and factor the greatest common divisor of the resulting matrix entries into irreducible factors. If one of these irreducible factors has a (prime) degree in the desired range, we choose it as polynomial $f(x)$ defining our field. Otherwise we can test another candidate for b. As shown in [SGGB00], this strategy to generate a trapdoor works quite nicely, using a ν-value as small as 3.

Solution to Exercise 62. The hash function h and the underlying compression function f operate on message blocks of length n. Hence, choosing sk_{in} and sk_{out} to be of length n enables a (pre-)processing of the first input block to h—before the message m is available.

A.4   Solutions to selected exercises of Part IV

Solution to Exercise 64. Participant U_i has been able to compute $\alpha^{P(2j-1)}$ for $j = 1, \ldots, m$, and $\alpha^{P(2i)}$. Now, P is a degree m polynomial over the field $\mathbb{Z}_q$, so it suffices to adapt Lagrange interpolation to the exponents. Given $m+1$

tuples $(x_0, P(x_0)), \ldots, (x_m, P(x_m))$, with pairwise distinct left components, we know

$$P(x) = \sum_{k=0}^{m} P(x_k) L_k(x)$$

where

$$L_k(x) = \prod_{\substack{t=0 \\ k \neq t}}^{m} \frac{x_t - x}{x_t - x_k}.$$

As a result,

$$P(0) = \sum_{k=0}^{m} P(x_k) L_k(0) = \sum_{k=0}^{m} P(x_k) \cdot \prod_{\substack{t=0 \\ k \neq t}}^{m} \frac{x_t}{x_t - x_k},$$

and hence

$$\alpha^{P(0)} = \alpha^{\sum_{k=0}^{m} P(x_k) \prod_{\substack{t=0 \\ k \neq t}}^{m} \frac{x_t}{x_t - x_k}} = \prod_{\substack{k=0 \\ k \neq t}}^{m} (\alpha^{P(x_k)})^{\prod_{\substack{t=0 \\ k \neq t}}^{m} \frac{x_t}{x_t - x_k}}.$$

Solution to Exercise 66. Indeed, the first equation states that $g^{\hat{a}} = g^a$. The further kn^2 equations guarantee that $\hat{a}$ is in $C(B)$, and, as a result, $g^{\hat{a}b} = g^{b\hat{a}} = g^{ab}$.

Solution to Exercise 68. The adversary may compute:

- $M_1 = P_1^{-1} D_1 P_1$ and $M_2 = P_2^{-1} D_2 P_2$ (D_1 and D_2 diagonal)
- $G = P_1 P_2^{-1}$
- $U = P_1 C P_2^{-1} = D_1^r G D_2^s$
- $V = P_1 D P_2^{-1} = D_1^v G D_2^w$

Setting $X = D_1^{r+v} G D_2^{s+w}$, it is clear that $K_{ab} = P_1^{-1} X P_2$, and therefore, if the adversary is capable of computing X, then the adversary can recover the whole matrix K_{AB}.

Solution to Exercise 70. The adversary may follow the following simple strategy: if two r-values in the signature of two different messages m_1 and m_2 coincide, let us assume they have been signed using the same k-value. Thus, given the message/signature pairs

$$(m_1, (s_1, r)) \text{ and } (m_2, (s_2, r))$$

the adversary notices

$$s_1 - s_2 = k^{-1}(H(m_1) - H(m_2)) \pmod{q}$$

and retrieves k computing it as

$$k = \frac{H(m_1) - H(m_2)}{s_1 - s_2} \bmod q.$$

Once k has been retrieved, it is straightforward to recover the prover's secret key x, just as

$$x = (s_1 k - H(m_1)) r^{-1} \bmod q.$$

Solution to Exercise 72. Assume the identification protocol is run n times, and let $m_1, \ldots, m_n$ be the challenges presented to the legitimate verifier, all in $\{0, 1\}^k$. Now, the adversary is presented with a challenge m^* by an honest verifier; his probability of outputting a valid signature of that value is bounded by the maximum of $n/2^k$ and P_{forge}, where P_{forge} is the probability that he outputs a valid signature on m^*. Such a signature would constitute a *selective forgery*. Indeed, an EUF-CMA scheme would suffice for this application, but even a weaker notion is sufficient, namely *selective unforgeability under known message attacks*.

Solution to Exercise 74. Given a signature and the public key, the image $\mathcal{H}(m)$ of the corresponding message m under $\mathcal{H}$ is uniquely determined, as f is a permutation. So unless a collision for $\mathcal{H}$ occurs, the message m is uniquely determined by the signature. The probability for a collision to occur is negligible—and a collision could actually enable a forgery in the sense of EUF-CMA. So the FDH signature satisfies this stronger notion of unforgeability.

References

[AAG99] I. Anshel, M. Anshel, and D. Goldfeld. An algebraic method for public-key cryptography. *Mathematical Research Letters*, 6:1–5, 1999.

[ACP09] M. Abdalla, C. Chevalier, and D. Pointcheval. Smooth projective hashing for conditionally extractable commitments. In S. Halevi, editor, *Advances in Cryptology – CRYPTO 2009*, volume 5677 of *Lecture Notes in Computer Science*, pages 671–689. International Association for Cryptologic Research, Springer, 2009.

[AD94] L.M. Adleman and J. DeMarrais. A subexponential algorithm for discrete logarithms over all finite fields. In D.R. Stinson, editor, *Advances in Cryptology – CRYPTO '93*, volume 773 of *Lecture Notes in Computer Science*, pages 147–158. Springer, 1994.

[AFFP11] M.R. Albrecht, P. Farshim, J.-C. Faugére, and L. Perret. Polly Cracker, revisited. In D.H. Lee and X. Wang, editors, *Advances in Cryptology – ASIACRYPT 2011*, volume 7073 of *Lecture Notes in Computer Science*, pages 179–196. International Association for Cryptologic Research, Springer, 2011.

[AKS04] M. Agrawal, N. Kayal, and N. Saxena. PRIMES is in P. *Annals of Mathematics*, 160(2):781–793, 2004.

[Art47] E. Artin. Theory of braids. *Annals of Mathematics*, 48:101–126, 1947.

[AS11] E. Andreeva and M. Stam. The symbiosis between collision and preimage resistance. In L. Chen, editor, *Cryptography and Coding 2011*, volume 7089 of *Lecture Notes in Computer Science*, pages 152–171. Springer, 2011.

[Asc04] M. Aschbacher. The status of the classification of the finite simple groups. *Notices of the AMS*, 51(7):736–740, August 2004.

[ATS03] P.J. Abisha, D.G. Thomas, and K.G. Subramanian. Public key cryptosystems based on free partially commutative monoids and groups. In T. Johansson and S. Maitra, editors, *Progress in Cryptology – INDOCRYPT 2003*, volume 2904 of *Lecture Notes in Computer Science*, pages 218–227. Springer, 2003.

[ATVZ09] R. Álvarez, L. Tortosa, J.F. Vicent, and A. Zamora. Analysis and design of a secure key exchange scheme. *Information Sciences*, 179(12):2014–2021, 2009.

[Bau93] G. Baumslag. *Topics in Combinatorial Group Theory*. Lectures in Mathematics. Birkhäuser, 1993.

[BCM09] S.R. Blackburn, C. Cid, and C. Mullan. Cryptanalysis of the MST_3 public key cryptosystem. *Journal of Mathematical Cryptology*, 3(4):321–338, January 2009.

[BCPQ01] E. Bresson, O. Chevassut, D. Pointcheval, and J.-J. Quisquater. Provably authenticated group Diffie–Hellman key exchange. In P. Samarati, editor, *Proceedings of the 8th ACM Conference on Computer and Communications Security (CCS-8)*, pages 255–264. ACM, 2001.

[BD95] M. Burmester and Y. Desmedt. A secure and efficient conference key distribution system (extended abstract). In A. De Santis, editor, *Advances in Cryptology – EUROCRYPT '94*, volume 950 of *Lecture Notes in Computer Science*, pages 275–286. Springer, 1995.

[BD05] M. Burmester and Y. Desmedt. A secure and scalable group key exchange system. *Information Processing Letters*, 94(3):137–143, May 2005.

[Bel97] M. Bellare. A note on negligible functions. *Journal of Cryptology*, 15:2002, 1997.

[BFN+11] G. Baumslag, N. Fazio, A. Nicolosi, V. Shpilrain, and W.E. Skeith III. Generalized learning problems with applications to non-commutative cryptography. In X. Boyen and X. Chen, editors, *Provable Security – ProvSec 2011*, volume 6980 of *Lecture Notes in Computer Science*, pages 324–339. Springer, 2011.

[BGGM07a] J.S. Birman, V. Gebhardt, and J. González-Meneses. Conjugacy in Garside groups I: Cyclings, powers and rigidity. *Groups, Geometry and Dynamics*, 1(3):221–279, 2007.

[BGGM07b] J.S. Birman, V. Gebhardt, and J. González-Meneses. Conjugacy in Garside groups I: Periodic braids. *Journal of Algebra*, 2:746–776, 2007.

[BGGM08] J.S. Birman, V. Gebhardt, and J. González-Meneses. Conjugacy in Garside groups II: Structure of the ultra summit set. *Groups, Geometry and Dynamics*, 2(1):16–31, 2008.

[BGS06] J.-M. Bohli, B. Glas, and R. Steinwandt. Towards provably secure group key agreement building on group theory. In P.Q. Nguyen, editor, *Progress in Cryptology – VIETCRYPT 2006*, volume 4341 of *Lecture Notes in Computer Science*, pages 322–336. Springer, 2006.

[BGVS07] J.-M. Bohli, M.I. González Vasco, and R. Steinwandt. Secure group key establishment revisited. *International Journal of Information Security*, 6(4):243–254, June 2007.

[BKL98] J. Birman, K.H. Ko, and S.J. Lee. A new solution to the word and conjugacy problems in the braid groups. *Advances in Mathematics*, 139:322–353, 1998.

[BKL+07] A. Bogdanov, L.R. Knudsen, G. Leander, C. Paar, A. Poschmann, M.J.B. Robshaw, Y. Seurin, and C. Vikkelsoe. PRESENT: An ultra-lightweight block cipher. In P. Paillier and I. Verbauwhede, editors, *Cryptographic Hardware and Embedded Systems – CHES 2007*, volume 4727 of *Lecture Notes in Computer Science*, pages 450–466. Springer, 2007.

[BM04] C. Boyd and A. Mathuria. *Protocols for Authentication and Key Establishment*. Springer, 2004.

[BMS06] J.-C. Birget, S.S. Magliveras, and M. Sramka. On public-key cryptosystems based on combinatorial group theory. *Tatra Mountains Mathematical Publications*, 33:137–148, 2006.

[BN08] M. Bellare and C. Namprempre. Authenticated encryption: Relations among notions and analysis of the generic composition paradigm. *Journal of Cryptology*, 21(4):469–491, October 2008.

[Bon99] J. Bonisoli. On collineation groups of finite planes. Lecture notes from the course *Finite Geometries and Their Applications*. Available at http://www.maths.qmul.ac.uk/~pjc/design/notes.html, 1999.

[BR93] M. Bellare and P. Rogaway. Random oracles are practical: A paradigm for designing efficient protocols. In *First ACM Conference on Computer and Communications Security*, pages 62–73. ACM, 1993.

[BR94a] M. Bellare and P. Rogaway. Entity authentication and key distribution. In D.R. Stinson, editor, *Advances in Cryp-*

tology – CRYPTO '93, volume 773 of *Lectures Notes in Computer Science*, pages 232–249. Springer, 1994.

[BR94b] M. Bellare and P. Rogaway. Optimal asymmetric encryption: How to encrypt with RSA. In A. De Santis, editor, *Advances in Cryptology – EUROCRYPT '94*, volume 950 of *Lecture Notes in Computer Science*, pages 92–111. Springer, 1994.

[BR96] M. Bellare and P. Rogaway. The exact security of digital signatures – How to sign with RSA and Rabin. In U. Maurer, editor, *Advances in Cryptology – EUROCRYPT '96*, volume 1070 of *Lecture Notes in Computer Science*, pages 399–416. Springer, 1996.

[BRS02] J. Black, P. Rogaway, and T. Shrimpton. Black-box analysis of the block-cipher-based hash-function constructions from PGV. In M. Yung, editor, *Advances in Cryptology – CRYPTO 2002*, volume 2442 of *Lecture Notes in Computer Science*, pages 320–335. Springer, 2002.

[BSGVM05] J.-M. Bohli, R. Steinwandt, M.I. González Vasco, and C. Martínez. Weak keys in MST_1. *Designs, Codes and Cryptography*, 37(3):509–524, 2005.

[BvL13] A. Blass and M. van Leeuwen. Is every pure set of permutations a group? Available at http://math.stackexchange.com/questions/369233/is-every-pure-set-of-permutations-a-group, April 2013. Replies to a question by user fgrieu from April 22, 2013.

[CCD+05] D. Catalano, R. Cramer, I. Damgard, G. Di Crescenzo, D. Pointcheval, and T. Takagi. *Contemporary Cryptology*. Advanced Courses in Mathematics CRM Barcelona. Birkhäuser Verlag, 2005.

[CG75] D. Coppersmith and E. Grossman. Generators for certain alternating groups with applications to cryptography. *SIAM Journal of Applied Mathematics*, 29(4), December 1975.

[CGH04] R. Canetti, O. Goldreich, and S. Halevi. The random oracle methodology, revisited. *Journal of the ACM*, 51(4):557–594, 2004.

[CJ03] J.H. Cheon and B. Jun. A polynomial time algorithm for the braid Diffie–Hellman conjugacy problem. In D. Boneh, editor, *Advances in Cryptology – CRYPTO 2003*, volume 2729 of *Lecture Notes in Computer Science*, pages 212–225. International Association for Cryptologic Research, Springer, 2003.

[CP05] R. Crandall and C. Pomerance. *Prime Numbers: A Computational Perspective*. Springer, second edition, 2005.

[CS98] R. Cramer and V. Shoup. A practical public key cryptosystem provably secure against adaptive chosen ciphertext attack. In H. Krawczyk, editor, *Advances in Cryptology – CRYPTO '98*, volume 1462 of *Lecture Notes in Computer Science*, pages 13–25. Springer, 1998.

[CS02] R. Cramer and V. Shoup. Universal hash proofs and a paradigm for adaptive chosen ciphertext secure public-key encryption. In L.R. Knudsen, editor, *Advances in Cryptology – EUROCRYPT 2002*, volume 2332 of *Lecture Notes in Computer Science*, pages 45–64. Springer, 2002. Extended version available at `http://homepages.cwi.nl/~cramer/`.

[ČvTMH01] V. Čanda, Tran van T., S. Magliveras, and T. Horváth. Symmetric block ciphers based on group bases. In D.R. Stinson and S. Tavares, editors, *Selected Areas in Cryptography – SAC 2000*, volume 2012 of *Lecture Notes in Computer Science*, pages 89–105. Springer, 2001.

[CW93] K.W. Campbell and M.J. Wiener. DES is not a group. In E.F. Brickell, editor, *Advances in Cryptology – CRYPTO '92*, volume 740 of *Lecture Notes in Computer Science*, pages 512–520. Springer, 1993.

[DDF+84] M. Davio, Y. Desmedt, M. Fosséprez, R. Govaerts, J. Hulsbosch, P. Neutjens, P. Piret, J.-J. Quisquater, J. Vandewalle, and P. Wouters. Analytical characteristics of the DES. In D. Chaum, editor, *Advanves in Cryptology – CRYPTO '83*, pages 171–202. Plenum Press, 1984.

[Deh97] P. Dehornoy. A fast method for comparing braids. *Advances in Mathematics*, 125:200–235, 1997.

[Deh04] P. Dehornoy. Braid-based cryptography. *Contemporary Mathematics*, 360:5–33, 2004.

[DH76] W. Diffie and M. E. Hellman. New directions in cryptography. *IEEE Transactions on Information Theory*, 22(6):644–654, November 1976.

[DR02] J. Daemen and V. Rijmen. *The Design of Rijndael. AES – The Advanced Encryption Standard*. Information Security and Cryptography. Springer, 2002.

[dVP06] F. Levy dit Vehel and L. Perret. On Wagner–Magyarik cryptosystem. In Ø. Ytrehus, editor, *Coding and Cryptography, International Workshop, WCC 2005*, volume 3969 of *Lecture Notes in Computer Science*, pages 316–329. Springer, 2006.

[Dwo01] M. Dworkin. Recommendation for Block Cipher Modes of Operation. Methods and Techniques. NIST Special Publication 800-38A, December 2001. Available at http://csrc.nist.gov/publications/nistpubs/800-38a/sp800-38a.pdf.

[Dwo05] Morris Dworkin. Recommendation for Block Cipher Modes of Operation: The CMAC Mode for Authentication. NIST Special Publication 800-38B, May 2005. Avilable at http://csrc.nist.gov/publications/nistpubs/800-38B/SP_800-38B.pdf.

[Dwo07] M. Dworkin. NIST Special Publication 800-38D, November 2007. Available at http://csrc.nist.gov/publications/nistpubs/800-38D/SP-800-38D.pdf.

[ECH+92] D. Epstein, J. Cannon, D. Holt, S. Levy, M. Paterson, and W. Thurston. *Word Processing in Groups*. Jones and Bartlett Publishers, Inc., 1992.

[EG83] S. Even and O. Goldreich. DES-like functions can generate the alternating group. *IEEE Transactions on Information Theory*, 29(6):863–865, November 1983.

[EHK+13] A. Escala, G. Herold, E. Kiltz, C. Ràfols, and J.L. Villar. An algebraic framework for Diffie–Hellman assumptions. In R. Canetti and J.A. Garay, editors, *Advances in Cryptology – CRYPTO 2013, Part II*, volume 8043 of *Lecture Notes in Computer Science*, pages 129–147. International Association for Cryptologic Research, Springer, 2013.

[ElG85] T. ElGamal. A public key cryptosystem and a signature scheme based on discrete logarithms. In G. R. Blakley and D. Chaum, editor, *Advances in Cryptology – CRYPTO '84*, volume 196 of *Lecture Notes in Computer Science*, pages 10–18. Springer, 1985.

[Ell97] J.H. Ellis. The history of non-secret encryption, 1997. Available at http://cryptocellar.web.cern.ch/cryptocellar/cesg/ellis.pdf.

[FFS88] U. Feige, A. Fiat, and A. Shamir. Zero-knowledge proofs of identity. *Journal of Cryptology*, 1(2):77–94, 1988.

[FK94]   Michael Fellows and Neal Koblitz. Combinatorial cryptosystems galore! *Contemporary. Math*, 168:51–61, 1994.

[FOPS01] E. Fujisaki, T. Okamoto, D. Pointcheval, and J. Stern. RSA-OAEP is secure under the RSA assumption. In J. Kilian, editor, *Advances in Cryptology – CRYPTO 2001*, volume 2139 of *Lecture Notes in Computer Science*, pages 260–274. Springer, 2001.

[FOPS04] E. Fujisaki, T. Okamoto, D. Pointcheval, and J. Stern. RSA-OAEP is secure under the RSA assumption. *Journal of Cryptology*, 17(2):81–104, 2004.

[Gar07]   D. Garber. Braid group cryptography. arXiv:0711.3491 [cs.CR], November 2007. Available at http://arxiv.org/abs/0711.3941.

[GGM13]   V. Gebhardt and J. González-Meneses. Generating random braids. *Journal of Combinatorial Theory. Series A.*, 120(1):111–128, 2013.

[GIMS11] M. Grassl, I. Ilić, S. Magliveras, and R. Steinwandt. Cryptanalysis of the Tillich–Zémor hash function. *Journal of Cryptology*, 24(1):148–156, 2011.

[GKT02]   D. Garber, S. Kaplan, and M. Teicher. A new algorithm for solving the word problem in braid groups. *Advances in Mathematics*, 167(1):142–159, 2002.

[GKT+06] D. Garber, S. Kaplan, M. Teicher, B. Tsaban, and U. Vishne. Length-based conjugacy search in the braid group. *Contemporary Mathematics*, 418:75–87, 2006.

[GL06]   R. Gennaro and Y. Lindell. A framework for password-based authenticated key exchange. *ACM Transactions on Information and System Security (TISSEC)*, 9(2):181–234, 2006.

[GM02]   R. Gennaro and D. Micciancio. Cryptanalysis of a pseudorandom generator based on braid groups. In L. Knudsen, editor, *Advances in Cryptology – EUROCRYPT 2002*, volume 2332 of *Lecture Notes in Computer Science*, pages 1–13. Springer, 2002.

[Gol99]   O. Goldreich. *Modern Cryptography, Probabilistic Proofs and Pseudorandomness*, volume 17 of *Algorithms and Combinatorics*. Springer, 1999.

[Gol01]   O. Goldreich. *Foundations of Cryptography – Volume 1 Basic Tools*. Cambridge University Press, 2001.

[Gol08] O. Goldreich. *Computational Complexity: A Conceptual Perspective.* Cambridge University Press, 2008.

[GV05] M.I. González Vasco. On the security of a group based public key cryptosystem. In *Proceedings of the Workshop on Mathematicals Problems and Techniques in Cryptology*, CRM Quaderns, pages 91–98. CRM, 2005.

[GVHMS04] M.I. González Vasco, D. Hofheinz, C. Martínez, and R. Steinwandt. On the security of two public key cryptosystems using non-abelian groups. *Designs, Codes and Cryptography*, 32:207–216, 2004.

[GVMSV05] M.I. González Vasco, C. Martínez, R. Steinwandt, and J.L. Villar. A new Cramer–Shoup like methodology for group based provably secure encryption schemes. In J. Kilian, editor, *Theory of Cryptography Conference – TCC 2005*, volume 3378 of *Lecture Notes in Computer Science*, pages 495–509. Springer, 2005.

[GVP07] M.I. González Vasco and D. Pérez. Attacking a public key cryptosystem based on tree replacement. *Discrete Applied Mathematics*, 155:61–67, 2007.

[GVPdPTD10] M.I. González Vasco, A.L. Pérez del Pozo, and P. Taborda Duarte. A note on the security of MST_3. *Designs, Codes and Cryptography*, 55(2–3):189–200, 2010.

[GVPdPTDV14] M.I. González Vasco, A.L. Perez del Pozo, P. Taborda Duarte, and J.L. Villar. Cryptanalysis of a key exchange scheme based on block matrices. *Information Sciences*, 276(0):319–331, 2014.

[GVRS03] M.I. González Vasco, M. Rötteler, and R. Steinwandt. On minimal length factorizations of finite groups. *Experimental Mathematics*, 12(1):1–12, 2003.

[GVS01] M.I. González Vasco and R. Steinwandt. Clouds over a public key cryptosystem based on Lyndon words. *Information Processing Letters*, 80:239–242, 2001.

[GVS04] M.I. González Vasco and R. Steinwandt. A reaction attack on a public key cryptosystem based on the word problem. *Applicable Algebra in Engineering, Communication and Computing*, 14(5):335–340, 2004.

[GVS06a] M.I. González Vasco and R. Steinwandt. Chosen ciphertext attacks as common vulnerability of some group-and polynomial-based encryption schemes. *Tatra Mountains Mathematical Publications*, 33:149–157, 2006.

[GVS06b] M.I. González Vasco and R. Steinwandt. Pitfalls in public key cryptosystems based on free partially commutative monoids and groups. *Applied Mathematics Letters*, 19(10):1037–1041, 2006.

[GVT06] M.I. González Vasco and P. Taborda. New steps towards secure word-problems based encryption schemes: Analysis of a recent proposal. In *Actas de la IX Reunión Española sobre Criptología y Seguridad de la Información*, pages 276–286, 2006.

[GVV08] M.I. González Vasco and J.L. Villar. In search of mathematical primitives for deriving universal projective hash families. *Applicable Algebra in Engineering, Communication and Computing*, 19(2):161–173, 2008.

[GZ91] M. Garzon and Y. Zalcstein. The complexity of Grigorchuk groups with application to cryptography. *Theoretical Computer Science*, 88:83–98, 1991.

[HEO05] D.F. Holt, B. Eick, and E.A. O'Brien. *Handbook of Computational Group Theory*. Discrete Mathematics and Its Applications. Chapman & Hall/CRC, Boca Raton, 2005.

[Her12] G. Herold. Polly Cracker, revisited, revisited. In M. Manulis M. Fischlin, J. Buchmann, editor, *Public Key Cryptography – PKC 2012*, volume 7293 of *Lecture Notes in Computer Science*, pages 17–33. International Association for Cryptologic Research, Springer, 2012.

[HGS99] C. Hall, I. Goldberg, and B. Schneier. Reaction attacks against several public-key cryptosystems. In V. Varadharajan and Y. Mu, editors, *Information and Communication Security – ICICS '99*, volume 1726 of *Lecture Notes in Computer Science*, pages 2–12, 1999.

[HK12] S. Halevi and Y. Kalai. Smooth projective hashing and two-message oblivious transfer. *Journal of Cryptology*, 25(1):158–193, 2012.

[HMvT94] T. Horváth, S.S. Magliveras, and Tran van T. A parallel permutation multiplier for a PGM crypto-chip. In Y.G. Desmedt, editor, *Advances in Cryptology – CRYPTO '94*, volume 839 of *Lecture Notes in Computer Science*, pages 108–113. Springer, 1994.

[Hol04] P.E. Holmes. On minimal factorisations of sporadic groups. *Experimental Mathematics*, 13(4):435–440, 2004.

[HS02] D. Hofheinz and R. Steinwandt. A practical attack on some braid group based cryptographic primitives. In Y.G. Desmedt, editor, *Public Key Cryptography – PKC 2003*, volume 2567 of *Lecture Notes in Computer Science*, pages 187–198. Springer, 2002.

[HT03] J. Hughes and A. Tannenbaum. Length-based attacks for certain group based encryption rewriting systems. arXiv:cs/0306032 [cs.CR], June 2003. Available at `http://arxiv.org/abs/cs.CR/0306032`.

[Hug02] J. Hughes. A Linear Algebraic Attack on the AAFG1 Braid Group Cryptosystem. In L. Batten and J. Seberry, editors, *Information Security and Privacy – ACISP 2002*, volume 2384 of *Lecture Notes in Computer Science*, pages 176–189. Springer, 2002.

[Hun89] T.W. Hungerford. *Algebra*. Graduate Texts in Mathematics. Springer, 1989.

[JS13] T. Jager and J. Schwenk. On the analysis of cryptographic assumptions in the generic ring model. *Journal of Cryptology*, 26(2):225–245, 2013.

[Kah80] D. Kahn. Codebreaking in World Wars I and II: The major successes and failures, their causes and their effects. *The Historical Journal*, 23(3):617–639, 1980.

[Kat10] J. Katz. *Digital Signatures*. Advances in Information Security. Springer, 2010.

[KBC97] H. Krawczyk, M. Bellare, and R. Canetti. Keyed-hashing for message authentication. Request for Comments: 2104, February 1997. Available at `http://tools.ietf.org/html/rfc2104`.

[KD04] K. Kurosawa and Y. Desmedt. A new paradigm of hybrid encryption scheme. In M. Franklin, editor, *Advances in Cryptology – CRYPTO 2004*, volume 3152 of *Lecture Notes in Computer Science*, pages 426–442. International Association for Cryptologic Research, Springer, 2004.

[Ker83] A. Kerckhoffs. La cryptographie militaire. *Journal des sciences militaires*, IX:5–38, Janvier 1883.

[KH06] K. Kurosawa and S.-H. Heng. The power of identification schemes. In M. Yung, editor, *Public Key Cryptography – PKC 2006*, volume 3958 of *Lecture Notes in Computer Science*, pages 364–377. Springer, 2006.

[KJRS86] B.S. Kaliski Jr., R.L. Rivest, and A.T. Sherman. Is DES a pure cipher? (Results of More Cyclic Experiments on DES). In H.C. Williams, editor, *Advances in Cryptology – CRYPTO '85*, volume 218 of *Lecture Notes in Computer Science*, pages 212–226. Springer, 1986.

[KL07] J. Katz and Y. Lindell. *Introduction to Modern Cryptography*. Chapman and Hall/CRC Press, 2007.

[KLC$^+$00] K.H. Ko, S.J. Lee, J.H. Cheon, J.W. Han, J.-S. Kang, and C. Park. New public-key cryptosystem using braid groups. In M. Bellare, editor, *Advances in Cryptology – CRYPTO 2000*, volume 1880 of *Lecture Notes in Computer Science*, pages 166–183. Springer, 2000.

[KLMT10] O. Kharlampovich, G. Lysënek, A.G. Myasnikov, and N.W.M. Touikan. The solvability problem for quadratic equations over free groups is NP-complete. *Theory of Computing Systems*, 47(1):250–258, July 2010.

[KM07] N. Koblitz and A. Menezes. Another Look at "Provable Security." *Journal of Cryptology*, 20(1):3–37, 2007.

[KR11] L.R. Knudsen and M. Robshaw. *The Block Cipher Companion*. Information Security and Cryptography. Springer, 2011.

[Kra02] D. Krammer. Braid groups are linear. *Annals of Mathematics*, 155:131–156, 2002.

[KY07] J. Katz and M. Yung. Scalable protocols for authenticated group key exchange. *Journal of Cryptology*, 20(1):85–113, 2007.

[KY13] A.A. Kamal and A.M. Youssef. Cryptanalysis of Alvarez et al. key exchange scheme. *Information Sciences*, 223(0):317–321, 2013.

[Len11] A.K. Lenstra. Integer factoring. In *Encyclopedia of Cryptography and Security*, pages 611–618. Springer, 2nd edition, 2011.

[LLH01] E. Lee, S.J. Lee, and S.G. Hahn. Pseudorandomness from braid groups. In J. Kilian, editor, *Advances in Cryptology – CRYPTO 2001*, volume 2139 of *Lecture Notes in Computer Science*, pages 486–502. Springer, 2001.

[LN97] R. Lidl and H. Niederreiter. *Finite Fields*, volume 20 of *Encyclopedia of Mathematics and Its Applications*. Cambridge University Press, 1997.

[LPPS07] G. Leander, C. Paar, A. Poschmann, and K. Schramm. New lightweight DES variants. In A. Biryukov, editor, *Fast Software Encryption, 14th International Workshop, FSE 2007*, volume 4593 of *Lecture Notes in Computer Science*, pages 196–210. International Association for Cryptologic Research, Springer, 2007.

[LS11] J. Lee and M. Stam. MJH: A faster alternative to MDC-2. In A. Kiayias, editor, *Topics in Cryptology – CT-RSA 2011*, volume 6558 of *Lecture Notes in Computer Science*, pages 213–236. Springer, 2011.

[LvT05] W. Lempken and Tran van T. On minimal logarithmic signatures of finite groups. *Experimental Mathematics*, 14(3):257–269, 2005.

[LvTMW09] W. Lempken, Tran van T., S.S. Magliveras, and W. Wei. A public key cryptosystem based on non-abelian finite groups. *Journal of Cryptology*, 22(1):62–74, 2009.

[Mat08] F. Matucci. Cryptanalysis of the Shpilrain–Ushakov protocol for Thompson's group. *Journal of Cryptology*, 21(3):458–468, 2008.

[MBG+93] A.J. Menezes, I.F. Blake, X.H. Gao, R.C. Mullin, S.A. Vanstone, and T. Yaghoobian. *Applications of Finite Fields*, volume 199 of *The Springer International Series in Engineering and Computer Science*. Springer, 1993.

[MH78] R.C. Merkle and M.E. Hellman. Hiding information and signatures in trapdoor knapsacks. *IEEE Transactions on Information Theory*, 24(5):525–530, 1978.

[Mil87] W. Miller. The Maximum order of an element of a finite symmetric group. *The American Mathematical Monthly*, 94(6):497–506, June–July 1987.

[MK99] K. Murasugi and B.I. Kurpita. *A Study of Braids*, volume 484 of *Mathematics and Its Aplications*. Kluwer, 1999.

[MM92] S.S. Magliveras and N.D. Memon. Algebraic properties of cryptosystem PGM. *Journal of Cryptology*, 5(3):167–183, 1992.

[MMO85] S.M. Matyas, C.H. Meyer, and J. Oseas. Generating strong one-way functions with cryptographic algorithm. *IBM Technical Disclosure Bulletin*, 27(10A):5658–5659, 1985.

[MPW94] S. Murphy, K. Paterson, and P. Wild. A weak cipher that generates the symmetric group. *Journal of Cryptology*, 7:61–65, 1994.

[MS90]   L. Mathew and R. Siromoney. A public key cryptosystem based on Lyndon words. *Information Processing Letters*, 35:33–36, 1990.

[MSvT02] S.S. Magliveras, D. Stinson, and Tran van T. New approaches to designing public key cryptosystems using one-way functions and trapdoors. *Journal of Cryptology*, 15:285–297, 2002.

[MU07]   A.D. Myasnikov and A. Ushakov. Length based attack and braid groups: cryptanalysis of Anshel–Anshel–Goldfeld key exchange protocol. In T. Okamoto and X. Wang, editors, *Public Key Cryptography – PKC 2007*, volume 4450 of *Lecture Notes on Computer Science*, pages 76–88. International Association for Cryptologic Research, Springer, 2007.

[MvOV96] A.J. Menezes, P.C. van Oorschot, and S.A. Vanstone. *Handbook of Applied Cryptography*. CRC Press, October 1996.

[MW97]   A. Menezes and Y.H. Wu. The discrete logarithm problem in $GL(n, q)$. *Ars Combinatoria*, 47:23–32, December 1997.

[NC00]   M.A. Nielsen and I.L. Chuang. *Quantum Computation and Quantum Information*. Cambridge Series on Information and the Natural Sciences. Cambridge University Press, 2000.

[Nic14] F. Nicolas. The information hiding homepage. Available at http://petitcolas.net/kerckhoffs/index.html, 2014.

[NIS94] NIST. Digital Signature Standards. Federal Information Processing Standards Publication 186-4, October 1994.

[NIS99] NIST. Data Encryption Standard (DES). Federal Information Processing Standards Publication 46-3, October 1999. Withdrawn on May 19, 2005.

[NIS01] NIST. Specification for the Advanced Encryption Standard (AES). Federal Information Processing Standards Publication 197, November 2001.

[NIS08] NIST. The Keyed-Hash Message Authentication Code (HMAC). Federal Information Processing Standards Publication – FIPS PUB 198-1, July 2008. Available at http://csrc.nist.gov/publications/fips/fips198-1/FIPS-198-1_final.pdf.

[NIS12] NIST. Secure Hash Standard (SHS). Federal Information Processing Standards Publication 180-4, March 2012. Available at http://csrc.nist.gov/publications/fips/fips180-4/fips-180-4.pdf.

[NIS14] NIST. SHA-3 Standard: Permutation-Based Hash and Extendable-Output Functions. Federal Information Processing Standards Publication – DRAFT FIPS PUB 202, May 2014. Available at `http://csrc.nist.gov/publications/drafts/fips-202/fips_202_draft.pdf`.

[NY89] M. Naor and M. Yung. Universal one-way hash functions and their cryptographic applications. In D.S. Johnson, editor, *ACM Symposium on Theory of Computing – STOC '89*, pages 33–43. ACM, 1989.

[Pai99] P. Paillier. Public-key cryptosystems based on composite degree residuosity classes. In J. Stern, editor, *Advances in Cryptology – EUROCRYPT 1999*, volume 1592 of *Lecture Notes in Computer Science*, pages 223–238. Springer, 1999.

[Pap01] R.S. Pappu. *Physical One-Way Functions*. PhD thesis, Massachusetts Institute of Technology, 2001.

[Pat99] K.G. Paterson. Imprimitive permutation groups and trapdoors in iterated block ciphers. In L. Knudsen, editor, *Fast Software Encryption – FSE'99*, volume 1636 of *Lecture Notes in Computer Science*, pages 201–214. Springer, 1999.

[PGV94] B. Preneel, R. Govaerts, and J. Vanderwalle. Hash functions based on block ciphers: a synthetic approach. In D.R. Stinson, editor, *Advances in Cryptology – CRYPTO '93*, volume 773 of *Lectures Notes in Computer Science*, pages 368–378. Springer, 1994.

[PL02] J. Pieprzyk and C. H. Li. Multiparty key agreement protocols. *Computers and Digital Techniques*, 147(4):229–236, 2002.

[PLQ07] C. Petit, K. Lauter, and J.-J. Quisquater. Cayley hashes: A class of efficient graph-based hash functions. Available at `http://perso.uclouvain.be/christophe.petit/files/Cayley.pdf`, 2007.

[PM97] H. Petersen and M. Michels. On signature schemes with threshold verification detecting malicious verifiers. In B. Christianson, B. Crispo, M. Lomas, and M. Roe, editors, *Security Protocols Workshop*, volume 1361 of *Lecture Notes in Computer Science*, pages 67–78. Springer, 1997.

[Poi05] D. Pointcheval. Provable security for public key schemes. In *Advanced Course on Contemporary Cryptology*, pages 133–189. Birkhäuser, 2005.

[PQ10] C. Petit and J.-J. Quisquater. Preimages for the Tillich–Zémor hash function. In A. Biryukov, G. Gong, and D.R. Stinson, editors, *Selected Areas in Cryptography – SAC 2010*, volume 6544 of *Lecture Notes in Computer Science*, pages 282–301. Springer, 2010.

[PQ13] C. Petit and J.-J. Quisquater. Rubik's for cryptographers. *Notices of the AMS*, pages 733–739, June/July 2013.

[Rab79] M.O. Rabin. Digitalized signatures and public-key functions as intractable as factorization. Technical Report, Massachusetts Institute of Technology, 1979.

[Rog06] P. Rogaway. Formalizing human ignorance. Collision-resistant hashing without the keys. In P.Q. Nguyen, editor, *Progress in Cryptology – VIETCRYPT 2006*, volume 4341 of *Lecture Notes in Computer Science*, pages 211–228. Springer, 2006.

[Ros02] K.H. Rosen. *Discrete Mathematics and Its Applications*. McGraw-Hill Higher Education, 5th edition, 2002.

[Rot95] J.J. Rotman. *An Introduction to the Theory of Groups*. Graduate Texts in Mathematics. Springer, 1995.

[RS04] P. Rogaway and T. Shrimpton. Cryptographic hash-function basics: Definitions, implications, and separations for preimage resistance, second-preimage resistance, and collision resistance. In B. Roy and W. Meier, editors, *Fast Software Encryption – FSE 2004*, volume 3017 of *Lecture Notes in Computer Science*, pages 371–388. International Association for Cryptologic Research, Springer, 2004.

[RSA78] R.L. Rivest, A. Shamir, and L. M. Adleman. A method for obtaining digital signatures and public-key cryptosystems. *Commununications of the ACM*, 21(2):120–126, 1978.

[Say94] M. Sayrafiezadeh. The birthday problem revisited. *Mathematics Magazine*, 67(3):220–223, June 1994.

[Sch04] B. Schneier. *Secrets and Lies – Digital Security in a Networked World. With new information about post-9/11 security*. Wiley, 2004.

[SDG06] H. Sibert, P. Dehornoy, and M. Girault. Entity authentication schemes using braid word reduction. *Discrete Applied Mathematics*, 154(2):420–436, 2006.

[SG02] R. Steinwandt and W. Geiselmann. Cryptanalysis of Polly Cracker. *IEEE Transctions on Information Theory*, 48(11):2990–2991, 2002.

[SGGB00] R. Steinwandt, M. Grassl, W. Geiselmann, and T. Beth. Weaknesses in the $SL_2(\mathbb{F}_{2^n})$ hashing scheme. In M. Bellare, editor, *Advances in Cryptology – CRYPTO 2000*, volume 1880 of *Lecture Notes in Computer Science*, pages 287–299. Springer, 2000.

[Sha49] C.E. Shannon. Communication theory of secrecy systems. *Bell Systems Technical Journal*, 28(4):656–715, 1949.

[Sha84] A. Shamir. A polynomial-time algorithm for breaking the basic Merkle–Hellman cryptosystem. *IEEE Transactions on Information Theory*, 30(5):699–704, 1984.

[Sho00] V. Shoup. Practical threshold signatures. In B. Preneel, editor, *Advances in Cryptology – EUROCRYPT 2000*, volume 1807 of *Lecture Notes in Computer Science*, pages 207–220. Springer, 2000.

[Sho01] V. Shoup. OAEP reconsidered. In J. Kilian, editor, *Advances in Cryptology – CRYPTO 2001*, volume 2139 of *Lecture Notes in Computer Science*, pages 239–259. Springer, 2001.

[Sho02] V. Shoup. OAEP reconsidered. *Journal of Cryptology*, 15(4):223–249, 2002.

[Sho04] V. Shoup. Sequences of games: A tool for taming complexity in security proofs. Cryptology ePrint Archive, Report 2004/332, November 2004. http://eprint.iacr.org/2004/332.

[Sho06] V. Shoup. *A Computational Introduction to Number Theory and Algebra*. Cambridge University Press, 2006.

[Shp92] I.E. Shparlinski. *Computational and Algorithmic Problems in Finite Fields*, volume 88 of *Mathematics and Its Applications*. Kluwer Academic, 1992.

[Shp99] I.E. Shparlinski. *Number Theoretic Methods in Cryptography. Complexity Lower Bounds*. Progress in Computer Science and Applied Logic. Springer, 1999.

[Shp08] V. Shpilrain. Cryptanalysis of Stickel's key exchange scheme. In E.A. Hirsch, A.A. Razborov, A. Semenov, and A. Slissenko, editors, *Computer Science – Theory and Applications – CSR 2008*, volume 5010 of *Lecture Notes in Computer Science*, pages 283–288. Springer, 2008.

[Sim94] C.C. Sims. *Computation with Finitely Presented Groups*. Encyclopedia of Mathematics and Its Applications. Cambridge University Press, 1994.

[Sin99] S. Singh. *The Code Book: The Evolution of Secrecy from Mary, Queen of Scots, to Quantum Cryptography.* Doubleday, New York, NY, USA, 1999.

[Sma03] N.P. Smart. *Cryptography: An Introduction.* McGraw-Hill Education, 2003.

[SSM10] N. Singhi, N. Singhi, and S.S. Magliveras. Minimal logarithmic signatures for finite groups of Lie type. *Designs, Codes and Cryptography,* 55(2–3):243–260, 2010.

[STAS02] S.C. Samuel, D.G. Thomas, P.J. Abisha, and K.G. Subramanian. Tree replacement and public key cryptosystem. In A. Menezes and P. Sarkar, editors, *Progress in Cryptology – INDOCRYPT 2002,* volume 2551 of *Lecture Notes in Computer Science,* pages 71–78. Springer, 2002.

[Sti05] E. Stickel. A new method for exchanging secret keys. In *International Conference on Information Technology and Applications – ICITA 2005,* pages 426–430. IEEE Computer Society, 2005.

[Sto98] A. Storjohann. An $O(n^3)$ algorithm for Frobenius normal form. In *International Symposium on Symbolic and Algebraic Computation – ISSAC'98,* pages 101–105. ACM, 1998.

[SU05] V. Shpilrain and A. Ushakov. Thompson's group and public key cryptography. In J. Ioannidis, A. Keromytis, and M. Yung, editors, *Applied Cryptography and Network Security – ACNS 2005,* volume 3531 of *Lecture Notes in Computer Science,* pages 151–163. Springer, 2005.

[SU06] V. Shpilrain and A. Ushakov. A new key exchange protocol based on the decomposition problem. *Contemporary Mathematics,* 418:161–167, 2006.

[SZ09] V. Shpilrain and G. Zapata. Using decision problems in public key cryptography. *Groups, Complexity, and Cryptology,* 1:33–49, 2009.

[TK05] Y. Tauman Kalai. Smooth projective hashing and two-message oblivious transfer. In R. Cramer, editor, *Advances in Cryptology – EUROCRYPT 2005,* volume 3494 of *Lecture Notes in Computer Science,* pages 78–95. International Association for Cryptologic Research, Springer, 2005.

[Tob02] C. Tobias. Security analysis of the MOR cryptosystem. In Y.G. Desmedt, editor, *Public Key Cryptography – PKC 2003,* volume 2567 of *Lecture Notes in Computer Science,* pages 175–186. Springer, 2002.

[Tsa13] B. Tsaban. Polynomial-time solutions of computational problems in noncommutative-algebraic cryptography. *Journal of Cryptology*, November 2013.

[TZ94] J.-P. Tillich and G. Zémor. Hashing with SL_2. In Y. Desmedt, editor, *Advances in Cryptology – CRYPTO '94*, volume 839 of *Lecture Notes in Computer Science*, pages 40–49. Springer, 1994.

[Vau06] S. Vaudenay. *A Classical Introduction to Cryptography: Applications for Communications Security*. Springer, 2006.

[Wer93] R. Wernsdorf. The one-round functions of the DES generate the alternating group. In R. A. Rueppel, editor, *Advances in Cryptology – EUROCRYPT '92*, volume 658 of *Lecture Notes in Computer Science*, pages 99–112. Springer, 1993.

[Wer02] R. Wernsdorf. The round functions of RIJNDAEL generate the alternating group. In J. Daemen and V. Rijmen, editors, *Fast Software Encryption – FSE 2002*, volume 2365 of *Lecture Notes in Computer Science*, pages 143–148. Springer, 2002.

[WM85] N.R. Wagner and M.R. Magyarik. A public key cryptosystem based on the word problem. In G. R. Blakley and D. Chaum, editor, *Advances in Cryptology – CRYPTO '84*, volume 196 of *Lecture Notes in Computer Science*, pages 19–36. Springer, 1985.

Index